The Mechanics of Morphogenesis: How Physics Shapes Biological Form

Sheena

Table of Contents

Chapter 1. Introduction and Background

Since the beginning of human history, people have been searching for answers to the mysteries of life. This inherent natural curiosity makes us different as human beings. This book is a step to lay a cornerstone for our understanding of how life is formed, from the perspective of mechanical engineering. More specifically, it is about how the shapes of living biological tissues and body forms are sculpted by mechanics when complex and diverse organisms develop from simple fertilized eggs. The creation of these shapes is called morphogenesis.

1.1 Introduction: morphogenesis and the gap in our understanding

Morphogenesis, "the generation of form," is from the Greek roots morph, where "morphê" means "shape" and "genesis" means "creation". It is the biological process of the generation of elaborate and diverse shapes of cells, tissues and body forms.[1] Morphogenesis is one of three fundamental aspects of developmental biology along with the control of tissue growth and patterning of cellular differentiation [1-3]. Besides controlling the organized spatial distribution of cells generated during embryonic development of an organism, morphogenesis also plays vital roles in mature organisms, in cultured cells, and inside tumor cell masses [3-6].

First, morphogenesis plays a central role in generating the diverse and functional organism forms [7-14]. During embryonic development, through morphogenesis, simple and unstructured

---

[1] Broadly speaking, the word "morphogenesis" can be used to "refer to many aspects of development," but when used strictly, as how it is used in this book, "it should mean the moulding of cells and tissues in to definite shapes," which was described in Waddington's book *Principles of Embryology* in 1956 [1].

tissues (Fig. 1.1a) can generate remarkable shapes and structures, such as strikingly the patterned fruit fly eye (Fig. 1.1b) [7], densely branched lungs (Fig. 1.1c) [8-10], systematically buckled guts (Fig. 1.1d) [11, 12], and complexly folded brain cortex (Fig. 1.1e) [13, 14]. Morphogenesis can also take place in a mature organism, such as in the maintenance of tissue homeostasis by stem cells or in the regeneration of damaged tissues.

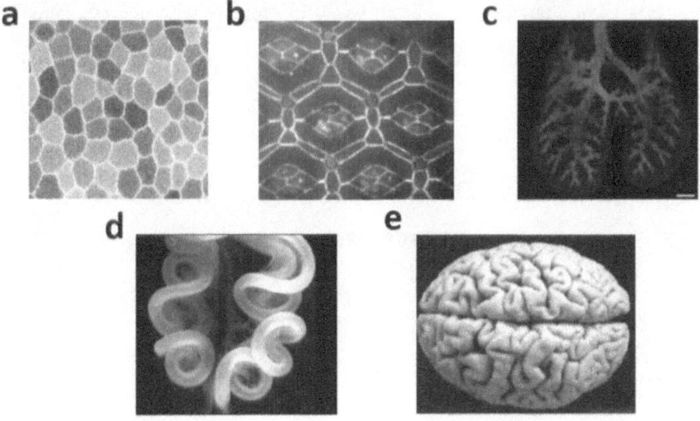

Figure 1.1. Morphogenesis generates diverse living tissue shapes. Through morphogenesis, simple tissues generate diverse and functioning tissues, organs, and body forms. (a) Segmented confocal image of epithelial tissue in the early *Drosophila* embryo. The tissue is relatively simple and unstructured at this time. (b) Confocal image of the *Drosophila* eye disc tissue expressing α-catenin::GFP at 28 hours after pupariation. Image from Lee *et al.*, 2010 (Image adapted with permission from Journal of Cell Science) [7]. (c) Optical projection tomography (OPT) image of whole-mount immunostained mouse lung (Image reprinted with permission from Nature Communications) [163]. (d) Bright-field image of looping of the chicken small intestine (Image reprinted with permission from the Nerurkar Lab, Columbia University) [11]. (e) Photograph of a formalin-fixed human brain taken from the dorsal side (Image reprinted with permission from American Association for the Advancement of Science) [203].

Second, understanding morphogenesis is also crucial to understanding and developing therapeutics for various diseases [15-23]. During human embryonic development, for example, if the neural tube tissue does not elongate and fold as it should to close completely, the baby may have a type of birth defect called Spina Bifida, which can cause physical and intellectual

disabilities (Fig. 1.2a) [15, 16]. In wounded animals, the skin tissue needs to remodel and change

shape to close the wound (Fig. 1.2b) [17, 18]. Another severe example of tissue morphogenesis is

the pathological case of cancer. Both cancer growth and metastasis are accompanied by highly

abnormal morphogenetic processes in tissues (Fig. 1.2d) [19, 20]. More recently, people have

observed morphological changes of the brain tissue in patients with neurodegenerative diseases

such as Alzheimer's Disease, which potentially sheds light on the understanding and treatment of

these diseases (Fig. 1.2c) [21-23].

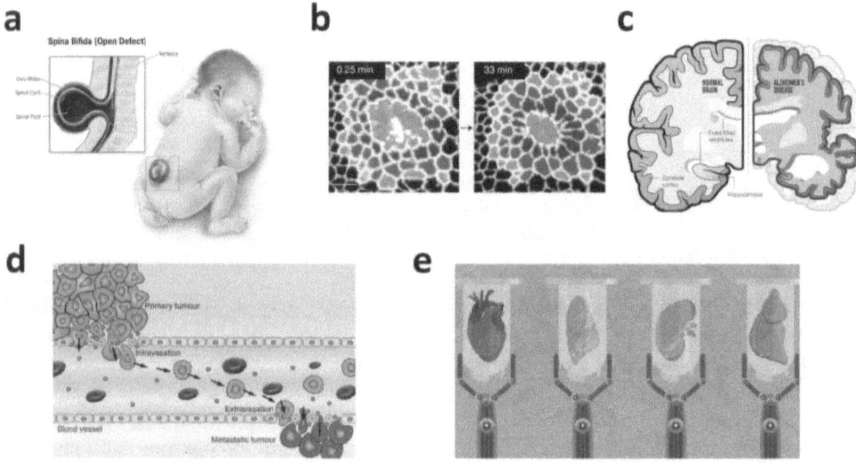

Figure 1.2. Morphogenesis and its applications in medicine and engineering. (a) Schematic of Spina Bifida, a birth defect due to incomplete enclosure of the neural tube. Image from Centers for Disease Control and Prevention–Beijing Medical University collaborative project [164]. (b) Color coded epithelial cells around a wound for time points immediately after wounding (left) and 33 min after (right). Cell membranes are labeled with GFP. Scale bar, 5 μm. Image from Tetley *et al.*, 2010 (Image reprinted by permission from Nature Physics) [18]. (c) Schematic of normal brains (left) and brains from patients with Alzheimer's Disease (right). Image from Drew, 2018 (Image adapted from illustration by Stacy Jannis/Alzheimer's Association) [165]. (d) Schematic of *in vivo* intravasation and extravasation of breast cancer cells by disrupting the blood vessel barrier. Image from Peng *et al.*, 2019 (Image reprinted by permission from Nature

Nanotechnology) [166]. (e) Cartoon of synthetic lab-grown human organs: test tubes with heart, kidney, lung and hepar. Image from ValentinaKru@ShutterStock [167].

Lastly, with the recent emergence of interdisciplinary studies combining engineering and biology, morphogenesis is being studied and controlled in the context of the creation and development of new forms in tissue engineering [24-26], multi-cellular engineered living systems [27-30], and microfluidic models such as organs-on-chips [31-33] (Fig. 1.2e).

The study of morphogenesis is in fact one of the oldest of all the sciences because many morphogenetic activities are observable [3]. In ancient Greece, Aristotle (384-322 BC) described structures in birds, mammals, fish and insects through observations and dissections (Fig. 3a) [34]. In the nineteenth century, one of the most famous studies of morphogenesis was conducted by Charles Darwin (1809-1882), who measured and described large numbers of species (Fig. 3b) and connected them to each other with his scientific theory of evolution [35]. At that time, however, the study of morphogenesis was limited to large-scale behaviors of organisms like organ and body forms. More recently, since the molecular revolution in biology started in the 1930s and the development of modern microscopes accelerated, increasing work has been done to understand molecular-scale activities in biology, such as the central dogma of biology, protein structure, signaling pathways etc. [3]. Many molecular signaling pathways that regulate morphogenesis have been revealed over the last several decades. For example, a series of work uncovered how cell behaviors are influenced by Rho GTPase proteins activating a number of downstream effectors, which in turn regulate cytoskeletal organization, intracellular trafficking and transcription (Fig. 3c) [36-39]. However, while researchers focused on these molecular studies, the large-scale physical and mechanical aspects of morphogenesis were largely ignored.

Therefore, many studies of morphogenesis have focused on the two extremes – either the very large or the very small length-scales. However, morphogenesis is a complex four-dimensional problem with three spatial dimensions and one temporal dimension involving the interaction of thousands of different molecules. There is a gap in our knowledge in connecting molecular-scale activities to larger-scale tissue and organ behaviors during morphogenesis (Fig. 3d). Even if all the molecules required for a process are known, we still do not understand how a functioning organism is made from these parts.

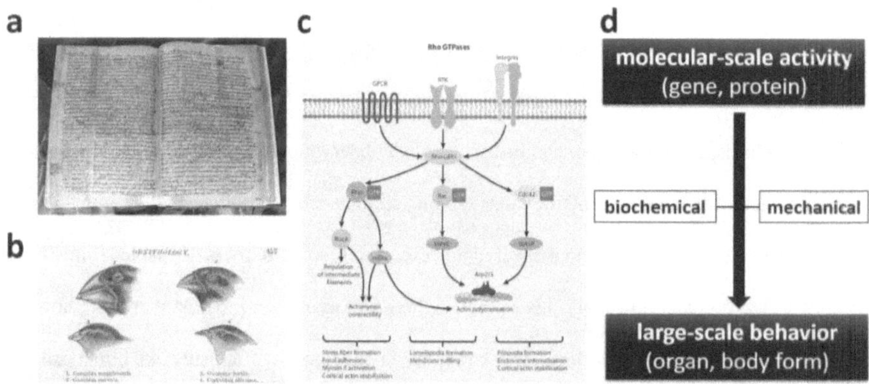

Figure 1.3. A brief history of morphogenesis research and the gap in our knowledge. (a) A page from the early version of *History of Animals*, one of Aristotle's writings on biology as the first in the history of science. (*Historia animalium et al.*, Constantinople, 12th century, Biblioteca Medicea Laurenziana, pluteo 87.4). About a quarter of his writings have survived. (b) Drawings of finches from the Galapagos Islands with beaks adapted to different diets from Charles Darwin's journals and notes. He collected bones and carcasses to bring back. Upon returning to London, he published his diaries as the "Journal of Researches Into the Geology and Natural History of the Various Countries Visited by H.M.S. Beagle." (c) Diagram showing cellular roles of Rho GTPases. Rho family members are key regulators of actin reorganization and intermediate filaments. Image from mechanobio.info [171]. (d) Diagram showing the gap in understanding morphogenesis. A complete understanding of both biochemical and mechanical factors involved in morphogenesis is needed to fill the gap connecting large-scale structures of animals like organ and body forms and molecular-scale activities like genes and proteins.

To fill this gap, a complete understanding of both the biochemical and mechanical factors involved in morphogenesis, from molecular to tissue length scales, is needed. Previous work has revealed significant insight on the biochemical side to elucidate the genes, proteins and chemical reactions needed for specific morphogenetic processes [40-42]. However, morphogenesis is naturally a mechanical process in which tissues are physically sculpted by mechanical stress, strain, and movements of cells that are induced by these genetic and molecular programs [3-6, 43-45]. Mechanics is the study of relationships among forces, matter and motion. We currently understand much less about the roles of mechanics in morphogenesis, such as how mechanical stresses and mechanical properties of tissues are regulated during these processes [43, 44, 46, 47]. This lack of understanding is in part because mechanical and biological factors are often strongly coupled together during morphogenesis, which makes these processes difficult to study, and because quantitative *in vivo* investigation methods are lacking in the field [6, 43-45].

To investigate mechanics during morphogenesis and to build a fundamental understanding of the mechanical mechanisms that generate tissue structures, this book explores mechanical factors in epithelial tissue morphogenesis during embryonic development of the model organism *Drosophila melanogaster*. This book focuses on epithelial tissues, within which cells are tightly packed and adhered to each other, due to their pivotal roles in shaping embryos and widespread presence in mature animals [48], and uses *Drosophila melanogaster*, a species of fruit fly, as a model organism because it is amenable to both genetic manipulation and high-resolution live imaging [48]. By developing and using systematic, quantitative, *in vivo* experimental approaches for studying epithelial tissue morphogenesis, this book addresses several important open questions in the field of the mechanics of morphogenesis: (1) what mechanical factors are involved in the morphogenesis of epithelial tissues; (2) how does a cell control these mechanical factors to tune tissue-level behaviors; (3) what applications can leverage these fundamental studies. This work was further strengthened by showing the capacity to expand the novel research approaches developed for the current two-dimensional studies of epithelial tissues to three-dimensions and to develop clinical applications by applying the experimental approaches above to other types of biological materials. A significant contribution of this book is that all the materials, methods, and tools have been developed with open-source resources and are freely available to the broader research community. These include fly stocks, protocols, programs, image processing tools, and experiment-verified models.

In this book chapter, Section 1.2 provides additional background, which is followed by the specific research objectives and the significance of this book in Section 1.3. This chapter closes with a detailed structure of this book in Section 1.4.

1.2 Background

This section introduces the current understanding of the roles of mechanics in morphogenesis emerging from both experimental and theoretical investigations, the powerful research techniques of confocal microscopy and quantitative image analysis, the structure and function of epithelial tissues, and the advantages of using *Drosophila melanogaster* as a model organism to study morphogenesis.

1.2.1 Mechanics in morphogenesis

Morphogenetic events are fundamentally mechanical in nature [43-47]. To comprehensively understand morphogenesis, we need to understand the relationships among the mechanical forces that drive tissue shape change, the mechanical properties that determine how tissues respond to forces, and the motions and shape changes that tissues undergo and maintain. With recent advances in molecular biology, microscopy, engineering methodology, and computational modeling, increasing work has been done to explore the roles of mechanics in morphogenesis both experimentally and computationally, highlighting the growing appreciation of the importance of mechanics during morphogenesis. However, few experimental studies of mechanics in morphogenesis during embryonic development have been conducted *in vivo*, especially during rapid morphogenetic events [43-47]. Moreover, even with the accelerating development of theoretical models to describe tissue morphogenetic behaviors, it is still quite challenging to connect parameters in models to physical results in experiments *in vivo*.

1.2.1.1 Experimental studies of mechanics in living tissues

Experimental studies of the last thirty years have begun to reveal mechanical mechanisms involved in specific morphogenetic processes. In the *Drosophila* eye (Fig. 1.1b), for example, the striking tissue pattern is believed to involve several different types of cell-cell adhesion mediated by three classical cadherins and the patterned cortical tension coupled with these adhesions [7, 49]. Buckling due to differential growth represents another mechanical mechanism driving morphogenetic events, such as the looping of the small intestine in chicken embryos (Fig. 1.1d) [11, 12]. Another stunning example of morphogenesis is branching in vertebrate lungs (Fig. 1.1 c). Recent work reveals that in different species, the mechanical mechanisms for branching are different. Branching in lungs is due to the active folding of the airway epithelium, the push from the lumen, or the push from the mesenchyme [8-10, 161, 162]. With the world's rapidly aging population and growing numbers of patients diagnosed with neurological diseases, recent work has increasingly started to focus on the brain. Recent experimental studies have illuminated not only the fundamental cellular and molecular processes underlying cortical brain development, but also the stress state, mechanical properties and spatiotemporal patterns of growth in the developing brain, indicating the key role of mechanics in brain development and functioning [13, 14, 22, 23,

50].

Despite progress in studying the mechanics of morphogenesis, few of these experimental studies have been conducted *in vivo* to directly measure tissue mechanical properties, especially during embryonic development [51-53]. Studies *in vivo* will be important for having full understanding of morphogenesis and to validate results for any clinical applications. However, *in vivo* measurements of mechanics in living tissues are challenging, in part because living tissues are usually small, soft, and difficult to be directly measured without destruction. Direct measurements of tissue mechanical properties and forces have been conducted on relatively large embryos like those of the model organism *Xenopus laevis*, which revealed soft and viscoelastic mechanical behavior (Fig. 1.4a) [54].

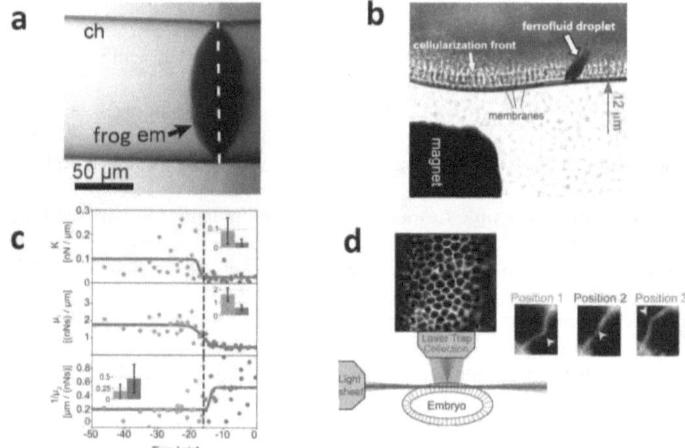

Figure 1.4. Prior experimental studies of mechanics in living tissues. (a) Micropipette aspiration experiment reveals viscoelastic epithelial tissue behavior of the frog embryo. Images from von Dassow *et al.*, 2008 (Image reprinted by permission from Developmental Dynamics) [54]. (b) A typical setup for probing tissue mechanical behavior in *Drosophila* embryos with ferromagnetic fluid droplets. A ferrofluid droplet that was injected in to the embryo is pulled by an external magnet through the epithelium. The red line shows the trajectory of the droplet. Images from Doubrovinski *et al.*, 2017 [55]. (c) A ferromagnetic fluid droplet experiment on epithelial tissues

in early *Drosophila* embryos reveals a switch in tissue mechanical properties during embryonic development. Red fitting curves show a step-like change of three mechanical property parameters during cellularization. Images from D'Angelo *et al.*, 2019 [56]. (d) Schematic of optical tweezer setup (left) and images of a cell interface in three different positions of deflection by an optical tweezer in early *Drosophila* embryos (right). Cell membrane is labeled by mCherry. Imaged from Bambardekar *et al.*, 2015 [57].

Direct *in vivo* mechanical measurements of *Drosophila* embryos have proven difficult as the embryo is small, delicate, and encased in an eggshell. Yet, some recent progress has been made [45, 55-57]. Several recent studies utilized magnetic manipulation of ferromagnetic droplets embedded inside early *Drosophila* embryos (Fig. 1.4b). By quantifying the droplet displacement and tissue deformation, one study demonstrated viscoelastic behavior of the epithelium in early *Drosophila* embryos, with predominantly viscous yolk cytoplasm and predominantly elastic cellular cortex [55]. Another study found a softening of epithelial tissues over time during development (Fig. 1.4c) [56]. In a different approach, optical tweezers and light-sheet microscopy were combined (Fig. 1.4d) to probe mechanical tensions at epithelial cell junctions, which were found to be on the order of 100 pN in early *Drosophila* embryos [57]. However, these measurements have focused on very early developmental stages when tissues are still immature, prior to the onset of morphogenetic movements. Moreover, these studies have not yet elucidated the molecular mechanisms that control tissue mechanical properties.

To address these gaps in our understanding, further studies will be needed that a) circumvent the many limitations faced in current *in vivo* experimental mechanical measurements and b) combine quantification of tissue mechanical behaviors and systematic manipulation of cell composition to reveal the nature and origin of the mechanical behavior of living tissues during morphogenesis.

1.2.1.2 Theoretical and computational modeling of mechanics in living tissues

Modeling plays a key role in building a comprehensive framework for understanding and predicting mechanical aspects of morphogenesis. Models can be used to develop abstract representations of morphogenetic events, test competing hypotheses in morphogenesis, and generate new predictions that can guide and be validated by experimental studies. There are several types of models that have been used to study epithelial tissue mechanics, such as vertex, particle jamming, cellular Potts, finite element and continuum models (Fig. 1.5) [58-61].

In self-propelled particle models, cells are treated as simple repulsive disks or spheres. These overdamped particles are driven by an active force to move and change direction by interactions with their neighbors or an external bath. This model has been used to explain jamming transitions in tissues, especially tissues with cellular density changes [61, 62] (Fig. 1.5e). The cellular Potts models describe cells as deformable objects with a certain volume that can adhere to each other and to the medium in which they live. It can be used to explain cell behaviors such as cell packing, sorting, migration and growth. Recently it has been used to explore how the balance between adhesion and tension influence cell shape and packing in the *Drosophila* retina [63] (Fig. 1.5f). Continuum mechanics approaches, including finite element models, describe tissues as a continuous mass rather than discrete particles (Fig. 1.5d and g). In finite element models, a large system is subdivided into smaller, simpler parts that are called finite elements. These models have been widely used to study tissue mechanics, such as bending, looping and remodeling. Recently, it has been successfully used to predict tissue flow during *Drosophila* ventral furrow invagination and body axis elongation [64, 65, 170] (Fig. 1.5g).

In the widely used vertex models of epithelial tissues, tissue shapes are quantified by a set of vertices and edges shared between adjacent cells and each cell is approximated geometrically by a polygon (2-D) or polyhedron (3-D). Cellular interfaces and cell areas/volumes can be defined from the positions of the vertices (Fig. 1.5a-c). Within the model, each vertex moves to reduce total tissue energy, which is defined as a function of cell vertex positions. This energy function corresponds to the work required to deform the junctional network of cells and is therefore determined by the energies associated with the shape of each cell. The vertex model has been intensively used to characterize cell packing topology, to study tissue growth, and to bridge the scales between force generation at the cellular level, and tissue deformation and flows [66]. This model focuses on the cell as a fundamental unit of the tissue and is primarily used in studying epithelial tissues with regular cell shapes such as polygons (2D) [67] (Fig. 1.5b and c) and prismoids (3D) [68]. Cell contractile tension and cell-cell adhesion are key parameters in vertex models (Fig. 1.5a). For example, a recent vertex model predicts that epithelial tissues can become more solid-like or more fluid-like depending on the balance between cell tension and cell-cell adhesion [67, 69, 70] (Fig. 1.5b). Remarkably, at the transition from unjammed to jammed behavior, cultured primary human bronchial epithelial cells take on shapes as predicted by the vertex model [71]. It has been proposed that developing tissues might behave like materials near a transition between solid-like and fluid-like behavior and that cells might regulate their properties by changing tension and/or adhesion [67, 69, 70].

With so many different types of models, however, it is still challenging to correctly connect models and experiments in studying the mechanical behaviors of living tissues. On the one hand, the coupling of mechanical and biological factors *in vivo* makes it difficult to either isolate specific physical parameters or consider the interactions between parameters in models. For example, although the recent vertex model successfully predicts the mechanical behaviors of cultured isotropic epithelial tissues [71], when it comes to *in vivo* epithelial tissues, many of which are highly anisotropic, the model fails to predict their mechanical behaviors [72]. On the other hand,

few *in vivo* experimental studies have been done to inform these models and test their predictions.

For example, the vertex model predicts that epithelial tissues would become more fluid-like with

increased cell-cell adhesion [67] (Fig. 1.5b), while a particle jamming model predicts more solid-

like behavior with increased cell-cell adhesion [40] (Fig. 1.5e). Quantitative experiments on how

cell-cell adhesion affects epithelial tissue mechanics *in vivo* are required to distinguish between

these models and to build a more precise and predictive experiment-modeling framework for

epithelial tissue mechanics.

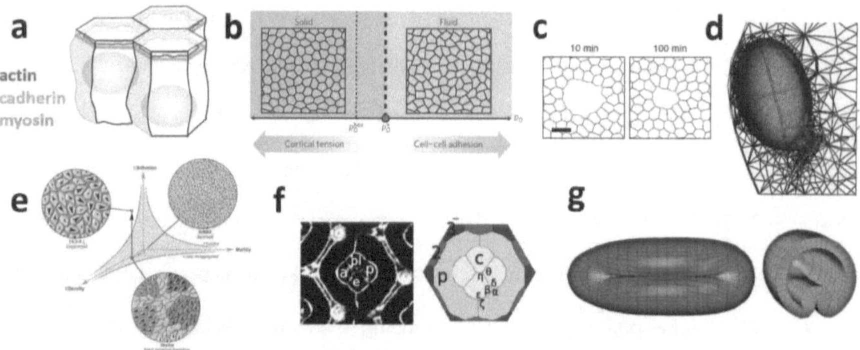

Figure 1.5. Different computational models for studying biomechanics in living tissues and related diseases. (a) Schematic of cells in an epithelial tissue. The balance of contractile tension generated by actin (red) and myosin (blue) and cell-cell adhesion mediated by cadherin (green) at the apical side of cells is thought to determine epithelial tissue mechanics. (Kong *et al.*, 2017, [47]). (b) The vertex model predicts that increased cell-cell adhesion would produce more fluid-like tissue behavior in continuous epithelium (Bi *et al.*, 2015, image reprinted by permission from Nature Physics [67]). (c) Vertex model simulation images of wound healing after ablation in epithelial tissues (Tetley *et al.*, 2019, image reprinted by permission from Nature Physics [18]). Scale bar, 5 μm. (d) Finite element analysis of uterus to study the mechanical environment of human pregnancy (Westervelt *et al.*, 2017, image reprinted by permission from American Society of Mechanical Engineers [169]). (e) The particle jamming model predicts that increased cell-cell adhesion would produce more solid-like tissue mechanical behavior when tissue densities can change (Sadati *et al.*, 2014, image reprinted by permission from WIREs Systems Biology and Medicine [61]). (f) The cellular Potts model (right) can be used to generate the same cell patterning

structure in the *Drosophila* retina ommatidium (left) (Käfer *et al.*, 2007 [63]). (g) A finite element model can be used to explore tissue invagination and bending in the *Drosophila* embryo (Conte *et al.*, 2008, image reprinted by permission from Journal of the Mechanical Behavior of Biomedical Materials [170]).

1.2.2 Microscopy and quantitative image analysis

Our understanding of the world rests on observation and experimental data. Although experimental sample materials vary in size, they are generally small in structure, especially in biological materials. The development of microscopes has broadened our ability to explore these objects that cannot been seen with the naked eye. A microscope is an instrument that magnifies objects, producing an image in which the object appears larger. The technical field of using microscopes to view samples and objects is called microscopy (Fig. 1.6a).

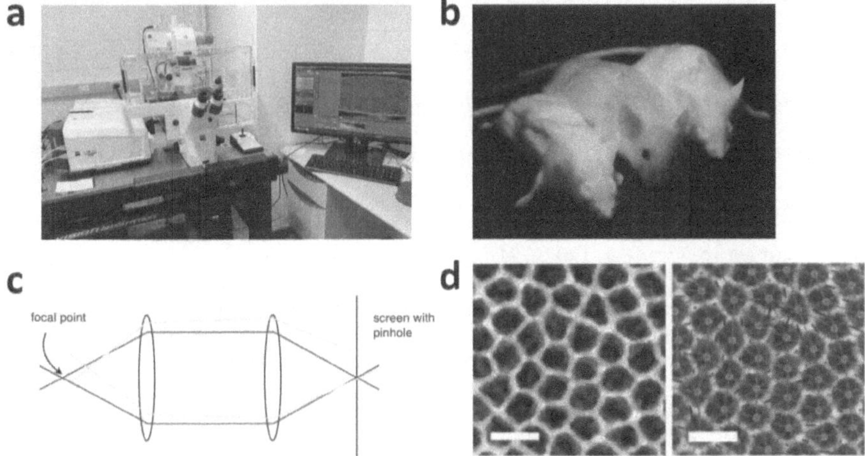

Figure 1.6. Microscopy and quantitative image analysis. (a) Picture of a confocal microscope (Zeiss LSM880 laser scanning confocal microscope). Image taken from the Kasza Lab at Columbia University. (b) Image of a trio of mice with two expressing the green fluorescent protein (GFP) gene under ultraviolet light. Image from Moen *et al.*, 2012 [174]. (c) A simplified view of the mechanism underlying confocal microscopy: rejection of light not incident from the focal plane.

All light from the focal point that reaches the screen is allowed through the pinhole. Light away from the focal point is mostly rejected. Image from Semwogerere *et al.*, 2005 [173]. (d) Segmentation of epithelial cells with the SEGGA software package (left) and quantification of cell shapes by the triangle method (right). Overlaid polygon representations (left) are used to quantify cell shapes. Cell centers (green dots) are connected with each other by a triangular network (red bonds) to quantify tissue anisotropy. Scale bar, 10 μm.

Among all types of microscopes, the optical microscope is most common; it uses lenses to refract visible light that passes through a thinly sectioned sample to produce an observable image. The first compound microscopes appeared around 1620 with one objective lens and one eyepiece lens. Just 50 years later, microscopes were already widely used to observe and examine small objects [5, 73]. Differential staining was discovered in the late 1850s, allowing a given cell structure to be distinguished from the rest in a fixed sample. Fluorescent dyes, compounds which emit light that has a longer wavelength after absorbing light with a shorter wavelength, have been used since then, and fluorescence microscopes, which uses fluorescence to generate an image, came into use. However, fluorescence microscopy was not widely used as a non-destructive imaging method until it was revolutionized by the introduction of the Green Fluorescent Protein (GFP) (Fig. 1.6b) as a tag of very low photo-toxicity in the 1990s [73, 75]. *Drosophila* became one of the first organisms whose structural and functional tissue behaviors were imaged in an almost non-invasive way (Fig. 1.6) [5, 74 181, 182].

Meanwhile, innovation in optical microscope mechanisms has progressed rapidly over the last century. Confocal microscopy, which increases optical resolution and contrast of a micrograph by using a spatial pinhole to block out-of-focus light in image formation, is already widely used in biology, medicine, and engineering research (Fig. 1.6a and c) [76]. Multi-photon excitation microscopy was invented later, which does not contain pinhole apertures that give confocal microscopes their optical sectioning quality. The optical sectioning produced by multiphoton microscopes is a result of the point spread function of the excitation, which can be a superior alternative to confocal microscopy due to its deeper tissue penetration, efficient light detection, and reduced photo-bleaching [77]. At the beginning of the 20th century, light-sheet fluorescence microscopy was introduced with intermediate-to-high optical resolution, good optical sectioning capabilities, and high speed. This microscopy often requires the samples to be chemically cleared [78].

Given the nature of the studies in this book on live and relatively accessible epithelial tissue structures *in vivo*, the imaging in this book is done by confocal fluorescence microscopy, unless otherwise specified.

Extracting information from image data is another challenge. With the rapidly increasing power of computing and programming, there are many emerging free and open-source toolsets that have been developed by the research community to analyze biological tissues. SEGGA, an image analysis software for automated image "SEGmentation, Graphical visualization and Analysis" developed by Dr. Jennifer Zallen's lab is an example (Fig 1.6d) [79]. This open-source software can be used to systematically track changes in cell shape, behavior and polarity in epithelial tissues and provides a suite of tools for fully automated image processing, image segmentation, cell tracking, data analysis and data visualization, as well as semi-automated error correction tools that expedite the process of obtaining accurate segmentation [79]. There are also similar platforms for biological tissue imaging analysis but utilizing different image segmentation and cell tracking algorithms, such as Tissue Analyzer from Dr. Suzanne Eaton's lab [80, 81, 130] and SIESTA from Dr. Rodrigo Fernandez-Gonzalez's lab [82, 83]. Very recently, with the development of artificial intelligence (AI), new methods combining machine learning and imaging processing for bioimage analysis are being developed. For example, EPySeg by Dr. Benjamin Prud'Homme's lab is focused on epithelial tissue image segmentation [84]. This open-source and coding-free software uses deep learning to segment membrane-stained epithelial tissues automatically and efficiently. It can be used as a Python package on a local computer or on the cloud for users not equipped with deep-learning compatible hardware. These new types of software packages have the potential to self-improve and to substantially reduce human input in image processing and analysis.

Using and building upon these open-source image analysis software platforms, quantitative analyses of molecular localization patterns, cell shapes, tissue structures, and tissue dynamics can be conducted on confocal image data. To support the research community, every method, algorithm, and protocol developed in this book is open-source and free to use.

1.2.3 Epithelial tissue and morphogenesis

During morphogenesis, the complex structures of multicellular organisms, such as tissues, organs and body forms, are generated from simple unstructured groups of cells [3, 4]. As one of the four basic types of animal tissue[2], epithelial tissues play vital roles in these events. Cells in epithelial tissues are often scutoid-shaped and adhere to one another to build almost impermeable tissue sheets organized in monolayers, as in developing embryos and the gut, or in multilayered as in the skin, with almost no intercellular spaces. For example, the early *Drosophila* embryo consists of a single-layer epithelial tissue sheet comprising columnar cells which are tightly associated with each other (Fig. 1.7a and b). Epithelial sheets act as barriers and boundaries to separate the inside of the body from the outside and different compartments of the body from each other (Fig. 1.7a) [3]. The functions of epithelial tissues require robust and cohesive tissue structures. However, even when cells are tightly packed, epithelial tissues can still be dynamic and display dramatic

---

[2] The other three basic animal tissue types are connective tissue, muscle tissue, and nervous tissue.

reorganization and remodeling, such as during the physical sculpting of embryonic tissues in many organisms [85]. It remains unclear how epithelial tissues accommodate dramatic shape changes while maintaining tissue integrity.

To connect cells and coordinate cell behaviors in epithelial tissues, there are four main types of cell-cell junctions: three types of connecting junctions that bind the cells together and one type of communicating junction [202]. The borders of two neighboring epithelial cells are often fused together around the whole perimeter of each cell, forming a continuous belt-like junction known as a tight junction or occluding junction. Below the tight junction, there usually lies adherens junctions, also known as anchoring junctions or zonula[3] adherens. Adherens junctions hold adjacent epithelial cells together by a protein called E-cadherin, which is further coupled to the intracellular actomyosin cytoskeleton. The coupled junctional actomyosin cytoskeleton tends to be arranged circumferentially around the cell as a marginal band, which can generate contractile tension to deform the shape of cells. Meanwhile, there is also actomyosin cytoskeleton concentrated at the medial apical cytoplasm, which can contract to change cell areas. Adherens junctions and their coupled actomyosin cytoskeleton are thought to be key in the morphogenesis of epithelial cells. At the basolateral part of epithelial cells, there are often desmosomes, also known as a spot desmosome or macula[4] adherens, connecting two cells together; and similar to desmosomes, there are hemidesmosomes at the basal part of epithelial cells to connect the basal surface of cells via intermediate filaments to the underlying basal lamina. There are also gap junctions between two adjacent cells to allow small molecules to pass between them.

Beyond their specialized cell-cell junction types, epithelial cells are often polarized with distinct structural orientations or protein localization patterns within a single cell. This polarity is

---

[3] In Latin, zonula means belt.
[4] In Latin, macula means spot.

characterized by cells with apical and basolateral membrane domains separated by adherens and tight junctions, and allows epithelial cells to transport molecules across the surface in a directional manner. The apical cell membrane often faces the lumen, while the basolateral membrane connects the cytoskeleton to extracellular matrix proteins within the basement membrane, which acts as a scaffolding on which the epithelium can grow and also serves as a selectively permeable membrane that determines which substances will be able to enter the epithelium. However, in some cases such as in early *Drosophila* embryos before gastrulation, there is little extracellular matrix at the basal side of the embryonic epithelium.

When epithelial tissues behave in a fluid-like and dynamic manner, such as during the physical shaping of embryos, the tissue sheets often undergo a movement called convergent extension that narrows and elongates tissues at the same time. Convergent extension is highly conserved and used in elongating tissues, tubular organs, and overall body shapes [86, 87, 121]. During body axis elongation of the *Drosophila* embryo (Fig. 1.7b), for example, the germband epithelial tissue rapidly doubles its length along the head-to-tail body axis in just 30 minutes. Convergent extension movements require anisotropic forces to drive tissue shape change, and an essential feature of many epithelia is anisotropy in the plane of the tissue sheet, known as *planar polarity*, which is associated with the asymmetric localization of key molecules inside cells [122-125]. For example, during *Drosophila* body axis elongation (Fig. 1.7b), the force-generating motor protein myosin II is planar-polarized. More specifically, myosin II molecules are specifically enriched at cell edges in the tissue that are oriented perpendicular to the head-to-tail body axis [40, 126], which is required for the tissue to rapidly elongate the body [64, 103, 119, 126-129].

Cell shape change and cell rearrangement are two key cellular mechanisms that contribute to epithelial tissue morphogenesis [41, 45, 89] (Fig. 1.7c). Epithelial cells can change shape without necessarily changing contacts with neighbors. They can also intercalate between each other, forming contacts with new cell neighbors during cell rearrangement. Coordinated cell rearrangement across a tissue can produce macroscopic changes in tissue dimensions [90]. For example, oriented cell rearrangement is thought to be the main driver of the rapid *Drosophila* embryonic body axis doubling in just 30 min (Fig. 1.7c) [41, 45].

Both cell shape and cell rearrangement in epithelial tissues are thought to depend on the balance between contractile tension in cells and cell-cell adhesion (Fig. 1.7d) [43, 91]. Contractile tension generated by the actomyosin cytoskeleton drives cells to change shape and rearrange, while cell-cell adhesion mediated by E-cadherin binds cells to each other and maintains tissue integrity. However, the dependence of epithelial tissue behaviors on contractile tension and cell-cell adhesion remains poorly defined.

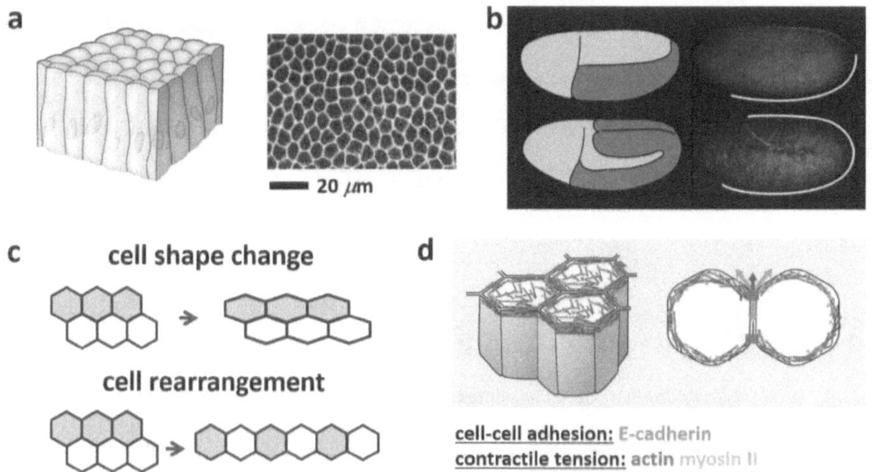

Figure 1.7. Epithelial tissue morphogenesis. (a) Schematic (left) and confocal image (right) of epithelial cells in the *Drosophila* embryo. Cell membranes are labeled with GFP. (b) Schematic (left) and confocal images (right) of body axis elongation in the *Drosophila* embryo. The germband tissue (dark gray area on the left, yellow line on the right) before (top) and after (bottom) the

epithelium narrows and elongates two-fold. Images from Kasza *et al*, 2010 (image adapted by permission from Current Opinion in Cell Biology) [89] and Fernandez-Gonzalez *et al*, 2009 (image adapted by permission from Developmental Cell) [128]. (c) Schematic of cellular mechanisms for epithelial tissue morphogenesis. Coordinated cell shape change (top) and cell rearrangement (bottom) drive epithelial tissue morphogenesis. (d) Schematic of molecular mechanisms for generating contractile tension and cell-cell adhesion in epithelial tissues. In the early *Drosophila* embryo, contractile tension is generated by actomyosin cytoskeleton and cell-cell adhesion is mediated by E-cadherin. Images modified from Heisenberg *et al*, 2013 (image adapted by permission from Cell) [43].

E-cadherin, a cell-cell adhesion molecule, plays a pivotal role in epithelial tissue behavior. The adhesive property of E-cadherin was first proposed in 1977 [92]. In *Drosophila*, the gene that encodes E-cadherin was named shotgun (*shg*) because mutating this gene results in many small holes in the epithelium, as if the tissue has been "shot" by a shotgun [93]. The importance of E-cadherin-based cell-cell adhesion for morphogenesis has been demonstrated in numerous systems [92-96]. However, much remains to be learned about the roles of cell-cell adhesion in epithelial tissue mechanics and morphogenesis. For example, it remains poorly understood how cell-cell adhesion influences tissue mechanical properties and how it couples with contractile tension to determine tissue mechanical behaviors. Systematic, quantitative experimental studies are needed to reveal how cell composition, including of the E-cadherin protein, governs epithelial tissue mechanics.

1.2.4 *Drosophila melanogaster* as a model organism

Drosophila melanogaster[5] is a species of fly in the family Drosophilidae, often referred to simply as the fruit fly (Fig. 1.8a). Since *Drosophila* was first used as a model organism by Charles William Woodworth (1865-1940) over a century ago, it remains one of the most studied organisms worldwide, particularly in genetics and developmental biology. Of particular note, Columbia

---

[5] For simplicity, *Drosophila melanogaster* is abbreviated as *Drosophila* in this book.

University played a central role in making *Drosophila* a major model organism in biological studies. The world-famous *Fly Room* was built by Dr. Thomas Hunt Morgan (1866-1945) in 1911 at Columbia University's Schermerhorn Hall (Fig. 1.8b), where Dr. Morgan demonstrated that genes are carried on chromosomes and are the physical basis of heredity. These discoveries formed the basis of the modern science of genetics [97, 98]. In fact, six[6] Nobel prizes have been awarded for research in *Drosophila* [98].

Drosophila has a short generation time of about 10 days and a lifespan of about 50 days at room temperature. The life cycle includes four stages: fertilized egg, larva, pupa, and adult (Fig. 1.8c). After being laid by the female adult, the half-millimeter long *Drosophila* embryo hatches into a larva within 24 hours at 25°C. The larva undergoes three stages before becoming a pupa (Fig. 1.8c), inside which the larval tissues are reabsorbed and the imaginal tissues undergo extensive morphogenetic movements to form adult structures [74, 99].

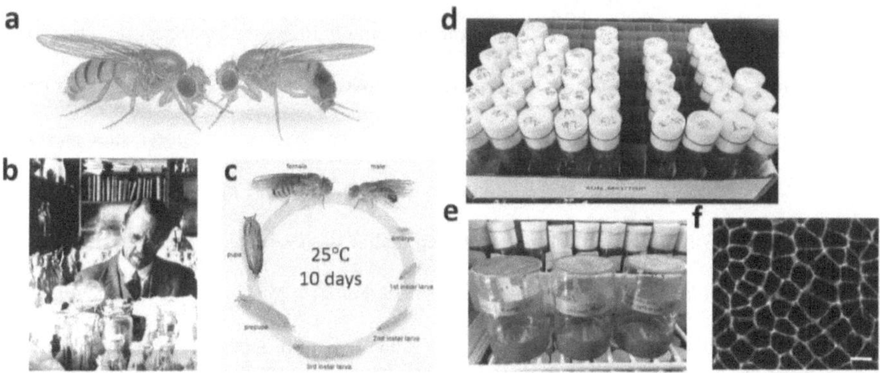

Figure 1.8. ***Drosophila melanogaster* as a model organism for studying mechanics in living tissues.** (a) Images of *Drosophila melanogaster* (fruit fly) female (left) and male (right). Images from California Institute of Technology, December 2016. (b) Picture of Dr. Thomas Hunter Morgan in the Fly Room at Columbia University with his large number of *Drosophila* cultures in 1922. Picture from Cal Tech Archives. (c) Images of the whole life cycle of *Drosophila*. The life cycle is relatively

____

[6] As of 2021.

rapid and takes only 10-12 days at 25 C. The *Drosophila* development is divided into four stages: embryo, larva (first instar, second instar and third instar), pupa and adult. Image from FlyMove database [172]. (d) *Drosophila* can be easily maintained in a lab in plastic vials with food. These vials take little equipment, space, and expense even when using large cultures. (e) Picture showing how *Drosophila* embryos are harvested to conduct experiment in a lab. *Drosophila* adults with different genotypes are kept in different plastic cups and the eggs are laid at the bottom of the cup on a plate of lab-made food. Embryos are collected from the plates. (f) A high resolution image of contractile and adhesive proteins in epithelial tissues in the *Drosophila* embryo. F-actin labeled by mCherry (red). E-cadherin labeled by GFP (green). Scale bar, 5 μm.

There are many reasons why *Drosophila* is a popular model organism. Besides its short life cycle, *Drosophila* also has a high fecundity, that is to say, a female can lay up to 100 eggs per day, and perhaps 2000 in a lifetime [99]. It is also very easy to maintain *Drosophila* in the lab with little equipment, space, and expense even when using large cultures (Fig. 1.8d). They can be safely and readily anesthetized simply by carbon dioxide gas or cooling in refrigerators. Moreover, fruit fly sex, virginity, and morphologies are all easy to identify to facilitate genetic crossing and studies [74].

In the context of studying mechanics during epithelial tissue morphogenesis, *Drosophila* is also an ideal model organism. First, there are many powerful genetic tools that we can use to manipulate genes and proteins in *Drosophila*. Second, the *Drosophila* embryo has accessible epithelial tissue sheets on which conventional confocal fluorescence microscopy studies can be conducted (Fig. 1.7a and b) [55]. Third, many epithelial tissues undergo morphogenesis at the surface of *Drosophila* embryos, which allow us to conduct high resolution live imaging and mechanical manipulation of the tissues (Fig. 1.8e and f). Fourth, morphogenetic movements in epithelial tissues in *Drosophila* embryos are rapid, which enables swift experimental progress [41, 89]. For example, during *Drosophila* embryo body axis elongation, the germband tissue doubles its length in just 30 minutes (Fig. 1.8). Finally, many biological and physiological properties are conserved between humans and *Drosophila*, and *Drosophila* shares 60% of human DNA and nearly 75% of human disease-causing gene functional homologs. Thus, studies in *Drosophila* have broad impacts on our understanding of human health and disease [100-102].

The majority of work in this book is conducted in *Drosophila* embryos.

1.3 Research significance and objectives

The research goal of this book is to advance our fundamental understanding of the mechanical mechanisms underlying how cells build and shape epithelial tissues with precise structural and mechanical properties that are required for proper biological function, which can also enhance our understanding of human development and human health.

Morphogenesis is the fundamental biological process that produces elaborate and diverse tissues and organs from simple groups of cells. This process controls the organized spatial distribution of cells generated during embryonic development and also plays roles in mature organisms, in cultured cells, and inside tumor cell masses [3-6]. One of the biggest challenges in understanding morphogenesis is the gap between our knowledge of the molecular-scale activities of genes and proteins, and the large-scale behaviors of tissues, organs and body forms [3-6]. Morphogenetic events are fundamentally mechanical in nature and there are significant mechanical factors [43, 44, 46, 47] involved in morphogenesis that remain poorly understood. A comprehensive understanding of the mechanical factors involved in development and morphogenesis, including both mechanical forces and tissue mechanical properties, is crucial to bridging our knowledge gap. During *Drosophila* embryonic development, for example, tissues dramatically change shape to lay out the basic plan for the body [40, 47, 89, 103, 104], but the underlying mechanical mechanisms driving these changes are not well understood. Although many genes and proteins required for tissue morphogenesis are known [40-42], we lack a complete understanding of how the activities of these molecules generate tissue structure and shape, in large part because 1) mechanical factors and biological factors are strongly coupled with each other during morphogenesis; 2) cells in living tissues actively generate forces and respond to the environment unlike passive forms of matter; 3) systematic, quantitative experimental methods to study the mechanics of epithelial tissues *in vivo* is lacking in the field.

In light of these challenges, this book a) combines experimental results and theoretical models of epithelial tissues to gain insight into the mechanics of morphogenesis by connecting tissue fluidity with cell shapes within the tissue; and b) applies systematic, quantitative, *in vivo* experimental approaches to advance fundamental understanding of the mechanics and functioning of epithelial tissues by revealing the role of cell-cell adhesion in epithelial tissue morphogenesis. These studies help fill the gaps in our understanding of morphogenesis by elucidating the role of mechanics in building functional tissue structures during morphogenesis, across molecular, cell, and tissue length scales.

To investigate mechanical factors during morphogenesis and to build a fundamental understanding of the mechanical mechanisms that generate tissue structure and function, the shape, movements and mechanics of epithelial tissues during *Drosophila* embryonic development were studied in this book. Theoretical models that can explain and predict *in vivo* epithelial tissue mechanical behaviors were developed and tested with modeling research collaborators. Furthermore, the experimental approaches developed to study epithelial tissue mechanics in two dimensions in *Drosophila* embryos in this book was expanded to explore epithelial mechanics in three dimensions and tissue mechanics in other types of biological materials such as the round window membrane in the inner ear of guinea pigs.

Specifically, this book includes three research objectives:

Research Objective 1: **To better understand and predict epithelial tissue mechanical behaviors** *in vivo***, I combined experimental studies and my collaborators' theoretical models to develop an experiment-modeling framework for understanding and predicting epithelial tissue mechanical properties.** Recently, both experimental studies and theoretical models have highlighted the linkage between tissue morphogenesis and tissue fluidity, which describes whether a tissue behaves in a solid-like or fluid-like manner [67, 71, 72, 105]. Fluid-like tissues accommodate tissue remodeling, while solid-like tissues resist tissue flow. However, how an epithelial tissue coordinates its fluidity to allow or prevent tissue reorganization at specific times remains unknown. I studied this question during a dramatic developmental event, in which the germband tissue rapidly elongates, doubling the body axis length of the *Drosophila* embryo. I used high-resolution confocal fluorescence imaging to take time-lapse movies of embryonic development, analyzed the shape and packings of cells in tissues, and combined experimental results with vertex model predictions from our theory collaborators to elucidate the role of tissue fluidity in this developmental event. We found that the fluidity of an anisotropic tissue can be read out by two experimentally accessible metrics of cell patterns within a tissue. This revealed that *Drosophila* body axis elongation is associated with a transition to more fluid-like tissue behavior in the germband epithelium. This work provides a new way to understand how mechanics influences morphogenesis, and an appealing approach to predict tissue behaviors simply by looking at a snapshot of cell shapes in the tissue. These findings may represent a general mechanical mechanism of morphogenesis in other tissue types and organisms, and this approach many be useful to engineers, physicists and biologists to probe mechanics in tissues where direct mechanical measurements are currently impossible, such as human embryonic tissues.

Research Objective 2: To explore the role of cell-cell adhesion in epithelial tissue structure and mechanics during morphogenetic events, I characterized how manipulating E-cadherin-based adhesion at cell-cell contacts influences epithelial tissue shape, structure, and movements during embryonic development *in vivo*. Many morphogenetic processes in epithelial tissues are thought to be determined by the balance of adhesion and tension [43, 91]. Adhesion physically connects cells to maintain tissue cohesion, but also organizes and transmits forces that promote tissue flow. It is unclear how adhesion enacts its unique dual functions in epithelial tissue morphogenesis. To address this, I used genetic approaches to tune cell adhesion levels in the *Drosophila* embryo, and combined live imaging and quantitative image analysis to study the effects of adhesion levels on cellular behaviors. I found biphasic dependencies of tissue mechanical behaviors on cell-cell adhesion levels, revealing surprising links between cell-cell adhesion, cell rearrangement, cell shape, and tissue fluidity. Furthermore, I showed that E-cadherin levels might contribute to these biphasic relationships by modifying adhesion dynamics and coupling to other molecules. This work is essential to understanding the roles of cell-cell adhesion during morphogenesis and development and developing a quantitative, multi-scale, experimental platform to explore mechanical factors *in vivo*. Using these approaches, I advanced fundamental understanding of how cell-scale adhesions control tissue-scale mechanical behaviors, which is essential to revealing mechanical mechanisms underlying epithelial tissue morphogenesis and embryonic development. Since many biological processes are conserved between humans and *Drosophila*, and nearly 75% of human disease-causing genes are believed to have a functional homolog in *Drosophila* [100-102], my work also sheds light on human embryonic development and how the improper regulation of tissue mechanics might be associated with birth defects, aberrant wound healing and cancer metastasis. These experimental results provide the necessary input for building and testing models of epithelial tissue mechanics and will be crucial for building tissues with precise shapes and structures in the lab.

Research Objective 3: **To more broadly elucidate the role of mechanics in tissue mechanics and morphogenesis, I expanded our experimental platform to further investigate how mechanics sculpts epithelial tissues in three dimensions and worked with collaborators to explore mechanics of membrane tissues to help develop new drug delivery methods into the inner ear.** Currently, most research on epithelial tissues in the early *Drosophila* embryo is focused on the apical side, where the majority of cortical tensions and adhesions junctions are localized. However, recent work has begun to show the importance of considering the tissue as a 3-D object [106-109]. The early *Drosophila* embryo is an ideal model to study the 3-D structure and mechanics of epithelia as the 5-10 µm wide columnar cells can grow to be about 30 µm tall along the apical-basal axis. By conducting deep live imaging and quantitative image analysis in the *Drosophila* embryo, I studied how the structure and mechanical behaviors of this epithelial tissue vary along each dimension. We found tissue structure and mechanical behaviors vary not only at the apical side over time, but also vary on different planes along the apical-basal axis. This work helps reveal the mechanical coupling between the apical and basal sides of tissues, elucidate the mechanics of morphogenetic events in three dimensions, and expand current tissue models from two to three dimensions.

The round window membrane (RWM) is a biological membrane that separates the fluid-filled inner ear from the air-filled middle ear. It plays an important role in modulating perilymphatic pressure and protecting the inner ear. While the integrity of the RWM is essential for normal hearing, it is also the only portal into the inner ear without requiring bone perforation [110, 111]. There are over 500 million people worldwide suffering from auditory and vestibular dysfunctions [110, 111]. Many of these diseases manifest themselves within the inner ear, such as Meniere's disease and sensorineural hearing loss. To access the inner ear directly for patient sampling and drug delivery, our collaborators proposed a novel approach by using ultra-sharp microneedles to perforate the RWM. In this collaboration, we first probed the RWM's structures and mechanical properties. I helped re-construct the membrane geometry and analyze orientations of different membrane fibers by using different imaging strategies. This is among the very first work to explore mechanics of this membrane. To develop a new method to deliver drugs into the inner ear, our collaborators fabricated an ultra-sharp needle to protrude the membrane to generate a recoverable hole to let drugs pass through. Using confocal imaging, I helped analyze the shape of the hole made by the needle, the change of tissue mechanics around the hole, and the recovery of the tissue following this perturbation.

In addition, our research community and diversity are important factors motivating me. There were many efforts I made throughout finishing this book to promote the inclusive and diverse environment of our lab, our department, our campus, and our community.

1.4 Structure of this book

In the following chapters, the mechanics of epithelial tissue morphogenesis during *Drosophila* embryonic development is studied and described as the three research objectives outlined in Section 1.3. In Chapter 2, a novel way to elucidate epithelial tissue morphogenesis and a new approach to understand and predict anisotropic epithelial tissue mechanical properties *in vivo* are developed. Combining methods developed in Chapter 2 with confocal microscopy, molecular biology, and quantitative image analysis, the role of cell-cell adhesion in regulating tissue mechanical behaviors *in vivo* is explored in Chapter 3. The study of epithelial tissue mechanics is further expanded to three dimensions, the mechanics of the round window membrane is analyzed by a series of collaborative projects, and a new drug delivery method into the inner ear is described in Chapter 4. A full discussion of the work produced in this book, its conclusions and significance, and suggestions for future research inspired by this work are presented in Chapter 5. All the related literature and publications can be found in the References chapter, and details of supplementary materials can be found in Appendix chapters.

Chapter 2. Epithelial Tissue Fluidity During Morphogenesis

Biological tissues dramatically change shape to form functional tissues and organs during embryonic development. Tissue flow and reorganization can happen on timescales as short as minutes. Epithelial tissues, for example, often narrow and elongate in convergent extension movements due to anisotropies in external forces or in internal cell-generated forces. However, the mechanisms that allow or prevent tissue reorganization, especially in the presence of strongly anisotropic forces, remain unclear. Tissue fluidity, which describes whether a developing tissue flows like a fluid or instead resists shape changes like a solid, has recently been identified as an important mechanism to control tissue behaviors during morphogenesis, although it is not well understood how mechanical and biological factors influence tissue fluidity. Combining experimental studies in the *Drosophila* embryo with modeling approaches, in Chapter 2, we show that the shapes and alignment of cells within tissues can help to elucidate and predict how tissues change their shapes and mechanical behaviors, including their fluidity, during development and how defects in these processes can result in abnormalities in embryo shape. Because many genes and cell behaviors are shared by *Drosophila* and humans, these results may reveal fundamental mechanisms underlying human development.

2.1 Introduction

2.1.1 Epithelial tissue morphogenesis and tissue fluidity

The ability of tissues to physically change shape and move is essential to fundamental morphogenetic processes that produce the diverse shapes and structures of tissues in multicellular organisms during development [112, 113]. Developing tissues are composed of cells that can dynamically change their behavior and actively generate forces to influence tissue reorganization and movement [45, 105, 114-117]. Remarkably, tissues dramatically deform and flow on timescales as short as minutes or as long as days [45]. Recently, both experimental studies and computational models highlight the linkage between tissue mechanical behaviors and the tissue fluidity: studies have shown that tissue movements within developing embryos can be linked with the tissue fluidity [18, 53, 105, 118], and computational models assuming predominantly fluid-like tissue behavior predict aspects of tissue movements [64, 119]. Fluid-like tissues accommodate tissue flow and remodeling, while solid-like tissues resist flow. Yet, the mechanisms underlying the mechanical behavior of developing tissues remain poorly understood, in part due to the challenges of sophisticated mechanical measurements inside embryos and the lack of unifying theoretical frameworks for the mechanics of multicellular tissues [45, 117, 120].

Epithelial tissue sheets play pivotal roles in physically shaping the embryos of many organisms [3, 4, 85, 113], often through convergent extension movements that narrow and elongate tissues at the same time. Convergent extension is highly conserved and used in elongating tissues, tubular organs, and overall body shapes [86, 87, 121]. Convergent-extension movements require anisotropies in either external forces that deform the tissue or asymmetries in cell behaviors that internally drive tissue-shape change. Indeed, an essential feature of many epithelia *in vivo* is anisotropy in the plane of the tissue sheet, a property known as planar polarity, which is associated with the asymmetric localization of key molecules inside cells [122-125]. For example, during *Drosophila* body axis elongation (Fig. 2.1a), the force-generating motor protein myosin II is specifically enriched at cell edges in the epithelial germband tissue that are oriented perpendicular to the head-to-tail body axis [40, 126] (Fig. 2.1b). Planar-polarized myosin is required for cell rearrangements that converge and extend the tissue to rapidly elongate the body and is thought to produce anisotropic tensions in the tissue [64, 103, 119, 126-129]. In addition, the *Drosophila* germband experiences external forces from neighboring tissues, including the mesoderm and endoderm, which have been linked to cell-shape changes in the germband during convergent extension [79, 131-133] (Fig. 2.1b). Meanwhile, the cells in the tissue are adhered to each other the whole time by adhesive forces from E-cadherin (Fig. 2.1b). Despite being fundamental to epithelial tissue behavior *in vivo*, it is unclear how such anisotropies arising from internal myosin planar polarity and external forces influence epithelial-tissue mechanical behaviors, particularly the tissue fluidity which determines whether the tissue behaves more like a fluid or a solid.

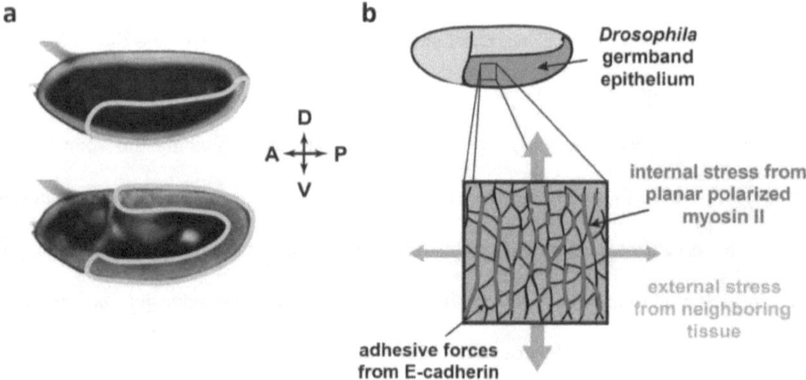

Figure 2.1. Mechanics in the converging and extending *Drosophila* germband epithelium during body axis elongation. (a) Bright-field images from time-lapse movies of *Drosophila* embryos undergoing body axis elongation. The germband epithelium (orange area) narrows and elongates (top to bottom) along the head-to-tail body axis (anterior to posterior) in a convergent extension movement. A, anterior; P, posterior; D, dorsal; V, ventral. (b) Schematic of *Drosophila* body axis elongation. The germband epithelium is shown in dark gray. The tissue is anisotropic, experiencing internal stresses from planar-polarized patterns of myosin II (red) within the tissue as well as external stresses (orange) due to the movements of neighboring tissue. Cells in the tissue are adhered to each other by adhesive forces from E-cadherin throughout this process.

2.1.2 Vertex models and tissue mechanical behaviors

Vertex models have proven a useful framework for theoretically studying the mechanical behavior of confluent epithelial tissues [66, 134], including the packings of cells in tissues [135-137] and the dynamics of remodeling tissues [81, 127, 135, 138, 139]. Recent studies of the energy barriers to cell rearrangement in isotropic vertex models, which assume no anisotropy in either internal tensions at cell–cell contacts or in external forces, have revealed a transition from solid to fluid behavior, which depends on whether large or small contacts are favored between neighboring cells. The transition is indicated by a single parameter describing cell shape, $\bar{p}$, which is the average value in the tissue of cell perimeter divided by the square root of cell area [67, 69, 70]. When cells prefer smaller contacts with neighbors, $\bar{p}$ is small, and the tissue is solid-like. Above a critical value of $\bar{p} = p_o^*$, the tissue becomes fluid-like. The isotropic vertex model successfully predicts that cell shapes identify the transition from fluid-like to solid-like behavior in cultured primary bronchial epithelial tissues; initial modeling work suggested that the critical cell shape p_o^* is close to 3.81 [67], in good agreement with the experiments [71]. Such a simple way to infer tissue behavior from static images is appealing, particularly for tissues that are inaccessible to mechanical measurements or live imaging.

However, subsequent work has shown that the precise value of p_o^* depends on specific features of the cell packing, such as the number of manyfold coordinated vertices [140] or the distribution of neighbor numbers in the packing [141-143], though the latter feature has never been studied systematically. In addition, these previous vertex model studies did not account for effects of anisotropy, potentially limiting their use in the study of converging and extending tissues.

2.1.3 Epithelial tissue mechanical behavior and tissue flow is linked to cell shapes and tissue anisotropy during convergent extension

In this chapter, we combine confocal imaging and quantitative image analysis with a vertex model of anisotropic tissues to study the converging and extending germband epithelium during *Drosophila* body axis elongation. We show that, in contrast to isotropic tissues, cell shape alone is not sufficient to predict the onset of rapid cell rearrangement during convergent extension in the *Drosophila* germband, which exhibits anisotropies arising from internal forces from planar-polarized myosin and external forces from neighboring tissue movements.

Combining experimental studies with theoretical considerations and vertex model simulations, we show that, for anisotropic tissues, such as the *Drosophila* germband, anisotropy shifts the predicted transition between solid-like and fluid-like behavior and so must be taken into account, which can be achieved by considering both cell shape and cell alignment in the tissue. We find that the onset of cell rearrangement and tissue flow during convergent extension in wild-type and mutant *Drosophila* embryos is more accurately described by a combination of cell shape and alignment than by cell shape alone. Moreover, we use experimentally accessible features of cell-neighbor relationships to quantify cell packing disorder and pinpoint p_o^*, which further improves our predictions. These findings suggest that convergent extension is associated with a transition from solid-like to more fluid-like tissue behavior, which may help to accommodate dramatic epithelial tissue-shape changes during rapid axis elongation. This study also provides a simple and appealing method to study tissue mechanical behaviors by just looking at snapshots of cells within the tissue, particularly useful for studying tissues where direct mechanical measurements or live imaging is inaccessible.

2.2 Materials and methods

The germband epithelial tissue is from *Drosophila* embryos generated at 23 °C and analyzed *in vivo* at room temperature. Cell outlines were visualized with gap43:mCherry [144], Spider:GFP, or Resille:GFP cell-membrane markers. Mechanical behavior of the epithelial tissue was altered in genetic mutants. Embryos were imaged on a Zeiss LSM880 laser-scanning confocal microscope. Time-lapse movies were analyzed with SEGGA software in MATLAB [79] for quantifying cell shapes and cell rearrangement rates, PIVlab (Version 1.41) in MATLAB [145] for quantifying tissue elongation, and custom code for quantifying cell alignment using the triangle method [130, 146, 147]. The vertex model describes an epithelial tissue as a planar tiling of N cellular polygons, where the degrees of freedom are the vertex positions [136]. Forces in the model were defined such that cell perimeters and areas act as effective springs with a preferred perimeter p_0 and a preferred area of one, which is implemented via an effective energy functional [143].

Unless otherwise noted, error bars are the standard deviations.

The data that support the findings of this study are included in Chapter 2 and Appendix A in this book. The custom code used in this study to extract the average triangle-based Q tensor from images segmented using SEGGA [79] is available at https://github.com/mmerkel/triangles-segga. Details of the materials and methods are listed below and detailed calculations can be found in Appendix A.

2.2.1 Experimental design

2.2.1.1 Fly stocks and genetics

The epithelial tissue studied was the germband tissue in *Drosophila* embryos. Embryos were generated at 23°C and analyzed at room temperature. Wild-type control embryos were *yw* with one maternal copy of a sqh-gap43:mCherry transgene to label cell membranes [144] (Fig. 2.2). *snail twist* embryos were zygotic *snail*[IIG05] *twist*[DfS60] mutants and expressed Spider:GFP to visualize cell outlines [79, 148]. The *bcd nos tsl* (*bnt*) maternal mutant embryos were the progeny of *bcd*[E1] *nos*[L7] *tsl*[146] homozygous females and expressed Resille:GFP to visualize cell outlines.

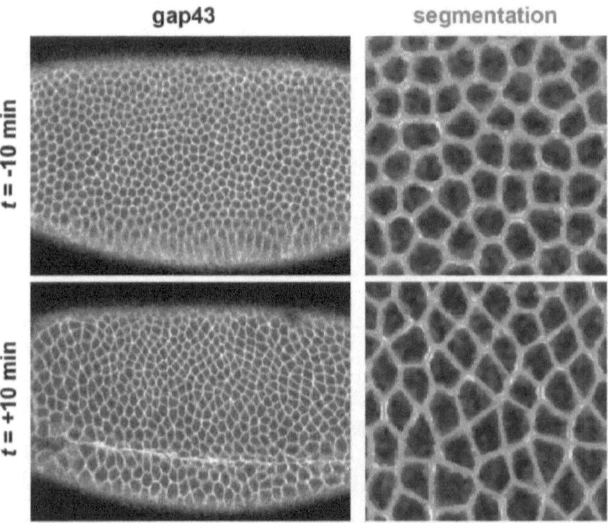

Figure 2.2. Imaging and analysis of the germband epithelial tissue. (*Left*) Confocal images from time lapse movies of epithelial cell patterns in the ventrolateral region of the germband tissue during *Drosophila* body axis elongation. Cell outlines were visualized using the fluorescently-tagged cell membrane marker, gap43:mCherry. Anterior left, ventral down. Images, 212 μm x 159 μm. (*Right*) Zoomed-in regions from images at left with overlaid polygon representations used to quantify cell shapes (green). Images, 40 μm x 40 μm.

2.2.1.2 Time-lapse imaging

Embryos aged 2-4 hours were dechorionated for 2 min in 50% (vol/vol) bleach, washed in distilled water, and mounted in halocarbon oil 27 and 700, 1:1 (Sigma) between a coverslip and an oxygen-permeable membrane (YSI). Embryos were oriented with the cephalic furrow and ventral furrow just visible at the edges of the field of view (Fig. 2.2). A 212 µm x 159 µm ventrolateral region of the embryo was imaged on a Zeiss LSM880 laser scanning confocal microscope with a 40X/1.2 NA water-immersion objective. Z-stacks were acquired at 1-µm steps and 15-s time intervals. Maximum intensity z-projections of 3 µm in the apical junctional plane (where the adhesions junctions are) were analyzed.

2.2.1.3 Tissue elongation measurements

Tissue elongation was measured by particle image velocimetry (PIV) using PIVlab in MATLAB [145]. Each image was divided into 2-pass Fast-Fourier-Transform windows (120 × 120 pixels) with 50% overlaps. A displacement vector field for each window and each time point was determined by cross-correlating each window in the current time point and the image in the next time point. Tissue length change was measured by quantifying the cumulative sum of the anterior-directed displacement at the anterior end of the germband and the posterior-directed displacement at the posterior end of the germband. The onset of tissue elongation ($t = 0$) was the time point when the derivative of the tissue elongation curve intersects zero (Fig. 2.4b).

2.2.1.4 Automated image segmentation and cell rearrangement analysis

Time-lapse movies were projected and despeckled using ImageJ [149]. Processed movies were segmented and computationally analyzed using the MATLAB based software SEGGA, and errors were corrected manually with the interactive user interface (Fig. 2.2) [79]. Cells were tracked and analyzed between $t = -10$ min and $t = 30$ min for each movie. To be included in cell rearrangement analysis, cells must be in the region of interest for at least 5 minutes after $t = 0$. The cell rearrangement rate shown is a uniformly weighted average over 1.5 minutes.

2.2.1.5 Cell shape index p and cell shape alignment Q analysis

Based on the cell segmentation data, we computed the average cell shape index $\bar{p}$ by quantifying for each segmented cell both the perimeter P and area A of the polygon defined by the cell vertices (i.e. the points where at least 3 cells meet). The average cell shape index $\bar{p}$ at a time point is the average of $P/\sqrt{A}$ over all segmented cells (Fig. 2.3a). The value of the cell shape index p is smallest when a cell takes a perfect round shape, where the value is 3.54. If the cell is stretched or has more edges, the value of p is increased. (Fig. 2.3a).

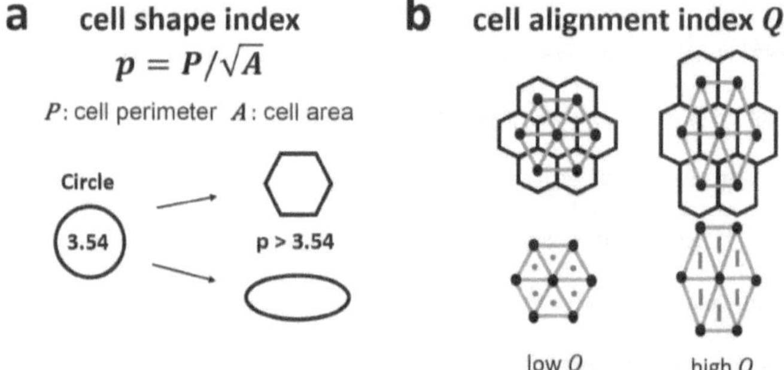

Figure 2.3. Cell patterns are quantified by the cell shape index p and the cell shape alignment Q. (a) p is the ratio of the cell perimeter P to the square root of cell area A. Circles have the smallest value of p = 3.54. (b) Q is quantified by the triangle method [72, 147] with a symmetric, traceless tensor q (red). Cell centers (black dots) are connected by a triangular network (blue bonds). Q is large when cells are stretched and aligned.

Cell shape alignment Q was quantified using the triangle method following Ref. [147]. A triangular tiling was created based on the barycenters of the cellular polygons (Fig 2.3b). For each triangle, we computed a symmetric, traceless tensor q quantifying triangle elongation. The cell shape alignment tensor Q is the average of the tensors q of all triangles. The cell shape alignment parameter Q in this book is the magnitude of the cell shape alignment tensor Q. This cell shape alignment parameter Q combines information about both cell shape anisotropy and cell shape alignment. Q is large only when the cell shapes are stretched and the stretches are aligned with each other (Fig2.3b). Details to calculate Q can be found in Appendix A.1.1.

2.2.2 Theoretical modeling

The theoretical modeling and related computational analyses were developed and done by collaborators in Lisa Manning's Group in the Department of Physics at Syracuse University.

2.2.2.1 Vertex model

Isotropic vertex model. The vertex model used describes an epithelial tissue as a planar tiling of N cellular polygons, where the degrees of freedom are the vertex positions [136]. Forces in the model were defined such that cell perimeters and areas act as effective springs with a preferred perimeter p_0 and a preferred area of one, which is implemented via an effective energy functional E. Periodic boundary conditions were used with box size $L_x \times L_y$ such that the average cell number density is one and the boundary conditions can accommodate a skew with a corresponding simple shear γ. We focused on stable, force-balanced states of the system, which corresponds to local minima of E. For a given local energy minimum, the simple shear modulus G was computed as described in [142]. Details to build the isotropic vertex model can be found in Appendix A.1.2.

Anisotropic vertex model. To introduce anisotropy into the isotropic vertex model, we did three sets of simulations. In all these simulations, the system was initialized with the Voronoi tessellation of a uniformly random point pattern on a squared domain. The effective energy was minimized after each step of introducing anisotropy.

For the first set of simulations of anisotropic tissue (Fig. 2.8a), an external pure shear strain ε was applied by setting $L_x = e^{\varepsilon} L_0$ and $L_y = e^{-\varepsilon} L_0$. For the second set of simulations (Fig. 2.8b,

Inset), the anisotropic myosin distribution in the germband was modeled by introducing an additional anisotropic line tension with amplitude λ_0 into the effective energy functional. In the second simulation, many states where the system could not reach any force-balanced state were found. In order to prevent the system from flowing indefinitely, a third set of simulations were run (Fig. 2.8b), where anisotropic line tensions were combined with a fixed system size. Details to build these anisotropic vertex models can be found in Appendix A.1.3.

Packing-dependence of the transition point. The precise values of the transition point p_o^* has proven to depend on specific features of the cell packing [140-143]. To study the packing-dependence of the transition point, we annealed the isotropic vertex model tissue at different temperatures prior to quenching the system to zero temperature. Results shown in Appendix A.2, Fig. A2, demonstrates that the transition point p_o^* does depend systematically on the annealing temperature and therefore on the packing disorder. Details about the packing-dependence of the transition point can be found in Appendix A.1.4 and Appendix A2 Fig. A2.

2.2.2.2 Theoretical expectation for the shift of the transition point

In a recent publication [143], our collaborators showed that the transition point $\overline{p}_{crit}$ in the vertex model is expected to shift away from the isotropic transition point p_0^* as the material is anisotropically deformed with strain ε as:

$$\overline{p}_{crit} = p_0^* + 4b\varepsilon^2 \tag{2.1}$$

where b is a constant prefactor whose precise value depends on the packing disorder. To compare this formula to the vertex model simulations, we need to take into account that cell rearrangements occur in our simulations. Removing their contribution from the overall tissue strain ε left us with only the cell shape alignment index Q (Appendix A.1.5), which can be quantified using methods in Section 2.2.1.5 and Appendix A.1.1. So we finally obtained $\overline{p}_{crit} = p_0^* + 4bQ^2$.

We then fit the new transition point to the simulation data from tissues under external anisotropic deformations (Fig. 2.8a, Appendix A.1.5.1, Appendix A.2 Fig. A3). To compare our theoretical predictions to experimental data, we define a quality of fit measure n_{tot}, which we define as the number of experimental data points that are wrongly categorized as either solid or fluid by our theory. To determine their value from experimental data, we minimize the quality of fit measure n_{tot} varying those fit parameters (Appendix A.1.5.2, Appendix A.2 Fig. A7).

Details about the shift of the transition point and its fit to simulation and experimental data can be found in Appendix A.1.5 and Appendix A.2 Fig. A2, A3.

2.3 Results

2.3.1 Cell shape alone is not sufficient to predict the onset of rapid cell rearrangement in the *Drosophila* germband epithelium

To explore the mechanical behavior of a converging and extending epithelial tissue in vivo, we investigated the *Drosophila* germband, a well-studied tissue that has internal anisotropies arising from planar-polarized myosin [40, 103, 104, 126-129] and also experiences external forces from neighboring developmental processes that stretch the tissue [131, 132]. The germband rapidly extends along the anterior–posterior (AP) axis while narrowing along the dorsal–ventral (DV) axis (Fig. 2.1), roughly doubling the length of the head-to-tail body axis in just 30 min [41] (Fig. 2.4b). Convergent extension in the *Drosophila* germband is driven by a combination of cell rearrangements and cell-shape changes (Fig. 2.4a, b). The dominant contribution is from cell rearrangement [41, 79, 103, 126], which requires a planar-polarized pattern of myosin localization across the tissue [40, 126] that is thought to be the driving force for rearrangement [104, 126-128]. Cell stretching along the AP axis also contributes to tissue elongation and coincides with movements of neighboring tissues [79, 131, 132, 151, 152], indicating that external forces play an important role in tissue behavior. Despite significant study of this tissue, a comprehensive framework for understanding its mechanical behavior is lacking, in part because direct mechanical measurements inside the *Drosophila* embryo, and more generally for epithelial tissues *in vivo*, continue to be a challenge [46, 51, 56].

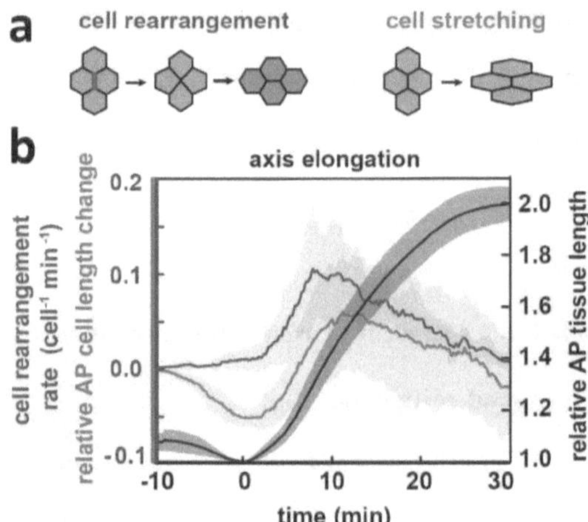

Figure 2.4. Cell shapes and cell rearrangements in the germband epithelium during *Drosophila* **body axis elongation.** (a) Schematic of oriented cell rearrangement and cell-shape change. (b) The germband epithelium doubles in length along the head-to-tail AP axis in 30 min (black). Cell rearrangements are thought to drive tissue elongation (magenta), and cell-shape changes also contribute (green). Tissue elongation begins at $t = 0$. The cell rearrangement rate includes cell-neighbor changes through T1 processes and higher-order rosette rearrangements. Relative cell length along the AP axis is normalized by the value at $t = -10$ min. Mean and SD between embryos is plotted ($n = 8$ embryos with an average of 306 cells analyzed per embryo per time point).

To gain insight into the origins of mechanical behavior in the *Drosophila* germband epithelium, we first tested the theoretical prediction of the vertex model that cell shapes can be linked to tissue mechanics. In the isotropic vertex model, tissue mechanical behavior is reflected in a single parameter, the average cell shape index $\bar{p}$ [67, 69-71]. To quantify cell shapes in the *Drosophila* germband, we used confocal time-lapse imaging of embryos with fluorescently tagged cell membranes [144] and segmented the resulting time-lapse movies [79] (Fig. 2.2 and Fig. 2.5a). Prior to the onset of tissue elongation, individual cells take on roughly isotropic shapes and become more elongated over time (Fig. 2.5a, b), consistent with previous observations [79, 131, 132, 153]. Ten minutes prior to tissue elongation, the cell shape index $\bar{p}$ averaged over eight wild-type embryos was just above 3.81. Eight minutes before the onset of tissue elongation, $\bar{p}$ started to increase before reaching a steady value of 3.98 about 20 min after the onset of tissue elongation (Fig. 2.5b). The average cell shape index prior to tissue elongation, $\bar{p} = 3.81$ (dashed line, Fig. 2.5b), was close to the value associated with isotropic solid-like tissues in previous work [67, 69, 70], suggesting that the tissue may be solid-like prior to elongation.

We next asked how these cell shapes vary among the individual embryos and correlate with tissue mechanical behavior. As an experimentally accessible read-out of tissue fluidity, we used the instantaneous rate of cell rearrangements occurring within the germband tissue (Fig. 2.4b), where higher rearrangement rates were associated with more fluid-like behavior and/or larger driving forces. Plotting instantaneous cell rearrangement rate versus $\bar{p}$ at each time point from movies of individual wild-type *Drosophila* embryos, we found that the onset of rapid cell rearrangement occurred at different values of $\bar{p}$ for each embryo, ranging from 3.83 to 3.90 for a cutoff rearrangement rate per cell of 0.02 min^{-1} (Fig. 2.5c). We verified that we observed a similar variation in the values of $\bar{p}$ for different cutoff values (Appendix A.2, Fig. A1). This suggests that, in the germband epithelium, comparing the cell shape index $\bar{p}$ to a fixed critical value (e.g., 3.81) is not sufficient to predict tissue behavior.

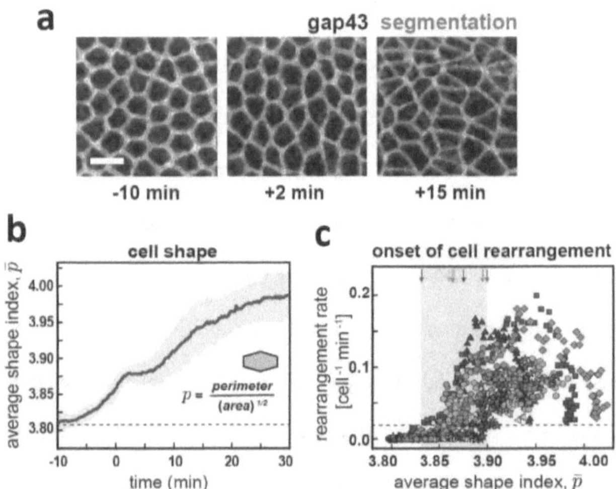

Figure 2.5. Cell shape alone is not sufficient to predict the onset of cell rearrangements in the *Drosophila* germband. (a) Confocal images from time-lapse movies of epithelial cell patterns in the ventrolateral region of the germband tissue during *Drosophila* axis elongation. Cell outlines were visualized by using the fluorescently tagged cell membrane marker gap43:mCherry [144]. Anterior left, ventral down. Images with overlaid polygon representations used to quantify cell shapes (green) are shown. (Scale bar, 10 μm.) See Fig. 2.2. (b) The average cell shape index $\bar{p}$ in the germband before and during convergent extension. The cell shape index, p, is calculated for each cell from the ratio of cell perimeter to square root of cell area, and the average value for cells in the tissue, $\bar{p}$, is calculated at each time point. The mean and SD between embryos is plotted. Dashed line denotes the reported value for the solid–fluid transition in the isotropic vertex model, $\bar{p} = 3.81$. See also Appendix A.2, Fig. A1. (c) The instantaneous rate of cell rearrangements per cell versus the average cell shape index $\bar{p}$ from movies of individual embryos at time points before and during convergent extension in eight wild-type embryos (different symbols correspond to different embryos). Small green arrows indicate the values of $\bar{p}$ at the onset of rapid cell rearrangement (>0.02 per cell per min; dashed line) in different embryos. The shaded region denotes values of $\bar{p}$ for which different embryos display distinct behaviors, either showing rapid cell rearrangement or not. Thus, a fixed value of $\bar{p}$ is not sufficient to determine the onset of rearrangement.

2.3.2 Cellular packing disorder is not sufficient to predict the onset of rapid cell rearrangement in the germband

Recent vertex model simulations suggest that $\bar{p} = 3.81$ is often insufficient to separate solid from fluid tissue behavior, as the precise location of the solid–fluid transition depends on how exactly cells are packed in the tissue [140-143]. A hexagonal packing has no packing disorder, while each cell with neighbor number different from six increases the packing disorder in the tissue. In the modeling literature, this disorder is typically generated either by allowing manyfold coordinated vertices (i.e., vertices at which more than three cells meet) or using simulation preparation protocols that create cell-neighbor numbers other than six. Including manyfold vertices in simulations is natural, as they are observed in the germband epithelium [153] and are often formed during cell rearrangements involving four or more cells [103, 126]. Moreover, recent theoretical work has predicted how the presence of manyfold vertices increases the critical shape index [140].

We wondered whether cell packing disorder quantified by the vertex coordination number z could explain the observed embryo-to-embryo variability in $\bar{p}$ at the transition point in wild-type embryos (Fig. 2.5c and Appendix A.2, Fig. A1). To test this idea, we plotted $\bar{p}$ versus z at each time point and color-coded the data based on the instantaneous cell rearrangement rate, pooling the data from all wild-type embryos (Fig. 2.6c). To isolate the changes in mechanical behavior of the germband during convergent extension from later developmental events, we focused on times $t \leq 20$ min after the onset of tissue elongation, well before cell divisions begin in the germband. If vertex coordination were sufficient to explain the germband behavior, then the theoretically determined line (dashed line) should separate regions with a low cell rearrangement rate (blue symbols) from regions with a high cell rearrangement rate (red, orange, and yellow symbols) (Fig.

2.6c). However, this was not the case, indicating that the prediction from ref. [140] alone is not sufficient to account for the germband behavior during this stage.

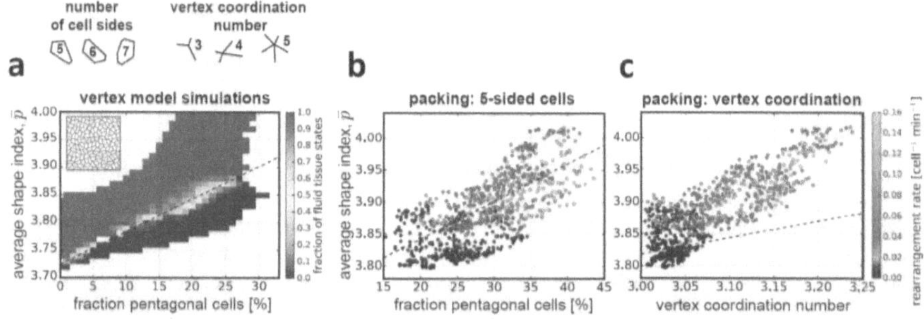

Figure 2.6. Cellular packing disorder alone is also not sufficient to predict the onset of cell rearrangements in the *Drosophila* germband. (a) In vertex model simulations, the solid–fluid transition depends on exactly how cells are packed in the tissue (Section 2.2 Materials and methods, Appendix A.1, Appendix A.2 Fig. A2). In model tissues, we find a linear dependence of the critical cell shape index on the fraction of pentagonal cells f_5, which is a metric for packing disorder. The dashed line represents a linear fit to this transition: $p_0^* = 3.725 + 0.59f_5$. (b) The relationship between $\bar{p}$ and f_5 for eight wild-type embryos, with each point representing a time point in a single embryo. The dashed line is the prediction from vertex model results (same as in a). (F) The relationship between $\bar{p}$ and vertex coordination number for eight wild-type embryos, with each point representing a time point in a single embryo. The dashed line is the prediction from ref. [140]. (b and c) Instantaneous cell rearrangment rate per cell in the tissue is represented by the color of each point, with blue indicating low rearrangement rates and red to yellow indicating high rearrangement rates.

Next, we asked if other aspects of packing disorder could affect tissue fluidity. Even without manyfold vertices, it is possible to generate packings *in silico* with differences in packing disorder just by altering the preparation protocol. Since this has not been systematically studied, we performed a large number of vertex model simulations where we varied the packing disorder (Appendix A.2, Fig. A2a and b). In our simulations, the transition point was well predicted by the fraction of pentagonal cells, i.e., cells that have exactly five neighbors, with a linear dependence (Fig. 2.6a and Appendix A.2, Fig. A2c and d). In particular, without any pentagonal cells, we recovered the predicted transition point of $\approx$3.72 for tissues consisting only of hexagonal cells [69, 136]. In comparison, the reported value of 3.81 corresponded to a fraction of $\approx$15% pentagonal cells (Fig. 2.6a). While additional aspects of cell packing likely affect the transition, these results suggest that the fraction of pentagonal cells may also be a good predictor for the transition point in isotropic tissues.

To test whether this second measure of packing disorder could explain the variability in $\bar{p}$ at the transition in wild-type embryos, we plotted $\bar{p}$ versus the fraction of pentagonal cells at each time point, again color-coding the data based on the instantaneous cell rearrangement rate and pooling data from all wild-type embryos (Fig. 2.6b and Appendix A.2, Fig. A1). We found that the packing disorder quantified by the fraction of pentagonal cells was also insufficient to explain the onset of cell rearrangements.

Our results suggest that two measures of packing disorder, the vertex coordination number and fraction of pentagonal cells, have at least partially independent effects on the isotropic vertex model transition point. However, neither of them is sufficient to understand the transition to high cell rearrangement rates in the *Drosophila* germband.

2.3.3 Theoretical considerations and vertex model simulations predict a shift of the solid–fluid transition in anisotropic tissues

To study whether anisotropies in the germband could affect the relation between the cell shape index and cell rearrangement rate, we used vertex model simulations to test how tissue anisotropy, introduced into the model in different ways, affects tissue fluidity.

First, we introduced anisotropy by applying an external deformation, mimicking the effects of forces exerted by neighboring morphogenetic processes, and then studied force-balanced states of the model tissue (Fig. 2.7a). As a metric for tissue stiffness, we measured the shear modulus of the model tissue, which describes with how much force a tissue resists changes in shape. A vanishing shear modulus corresponds to fluid behavior, where the tissue flows and cells rearrange in response to any driving force, whereas a positive shear modulus indicates solid behavior, where the tissue does not flow so long as the driving force is not too large.

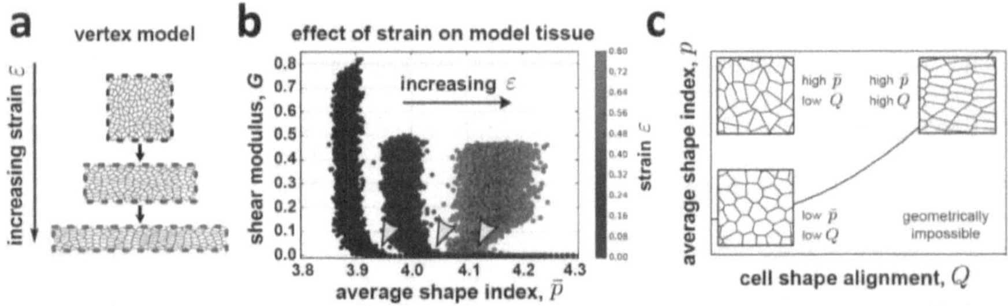

Figure 2.7. The vertex model of anisotropic tissues. (a) We study the effect of anisotropies on the solid–fluid transition in the vertex model by externally applying an anisotropic strain ε. An initially quadratic periodic box with dimensions $L_0 \times L_0$ is deformed into a box with dimensions $e^{\varepsilon}L_0 \times e^{-\varepsilon}L_0$. (b) Vertex model tissue rigidity as a function of the average cell shape index with different levels of externally applied strain ε (values for ε, increasing from blue to red: 0, 0.4, and 0.8). For comparison, the strain in the wild-type germband between the times $t = 0$ min and $t = 20$ min is $\varepsilon \sim 0.6$. For every force-balanced configuration, the shear modulus was analytically computed as described in Appendix A.1, Supplementary materials and methods. For zero strain, we find a transition at an average cell shape index of $\bar{p} = 3.94$ from solid behavior to fluid behavior. For increasing strain, the transition from solid to fluid behavior (i.e., the shear modulus becomes zero for a given strain) occurs at higher $\bar{p}$ (approximate positions marked by yellow arrows). Thus, a single critical cell shape index is not sufficient to determine the solid–fluid transition in an anisotropic tissue. (c) Cell shape and cell shape alignment can be used to characterize cell patterns in anisotropic tissues. Cell shape alignment Q characterizes both cell shape anisotropy and cell shape alignment across the tissue. While a high cell shape index $\bar{p}$ correlates with anisotropic cell shapes, the cell shape alignment Q is only high if these cells are also aligned. Conversely, low $\bar{p}$ implies low cell shape anisotropy and, thus, low Q.

We then analyzed how the shear modulus correlates with $\bar{p}$ for different amounts of global tissue deformation, quantified by the strain ε (Fig. 2.7b). For small strain, we recovered the behavior of the isotropic vertex model. The shear modulus was finite when $\bar{p}$ was small and vanished above a critical cell shape index, which was $p_0^* = 3.94$ for our simulations (Fig. 2.7b, blue symbols). For larger strains, we found that the critical value of the shape index at the transition between solid-like and fluid-like behavior generally increased with the amount of strain (Fig. 2.7b). Indeed, $\bar{p}$ for cells in a deformed, solid tissue can be higher than for cells in an undeformed, fluid tissue. This suggests that anisotropy affects the critical shape index at which the tissue transitions between solid and fluid behavior.

Some of us recently developed a theoretical understanding for a shift in the critical shape index when deforming a vertex model tissue [143]. In the limit of small deformations by some strain ε and without cell rearrangements, the critical value of $\bar{p}$ increases from p_0^* to $p_0^* + b\varepsilon^2$, where b is a constant prefactor. To compare this formula to the vertex model simulations (Fig. 2.7a and b), we need to take into account that cell rearrangements occur in our simulations. Removing their contribution from the overall tissue strain ε left us with a parameter Q (Fig. 2.7c) (Appendix A.1, Supplementary materials and methods), which can be quantified using a triangulation of the tissue created from the positions of cell centers (Appendix A.1, Supplementary materials and methods) [146, 147]. We term Q a "shape alignment index," as Q is nonzero only when the long axes of cells are aligned. We emphasize that, unlike the nematic-order parameter for liquid crystals, the cell alignment parameter Q is additionally modulated by the degree of cell shape anisotropy; tissues with the same degree of cell alignment but more elongated cells have a higher Q (Fig. 2.7c). In other words, Q can be regarded as a measure for tissue anisotropy. After accounting for cell rearrangements, we expect the transition point in anisotropic tissues to shift from the isotropic value p_0^* to (Appendix A.1, Supplementary materials and methods):

$$\overline{p}_{crit} = p_0^* + 4bQ^2 \tag{2.2}$$

Indeed, comparing this equation to vertex model simulations yields a good fit with the simulation results (Fig. 2.8a, solid line), with fit parameters $p_0^* = 3.94$ and $b = 0.43$ (Appendix A.2, Fig. A3). We confirmed that cell-area variation did not significantly affect these findings (Appendix A.2, Fig. A4). In principle, we expect both the transition point p_0^* and the precise value of b to depend on the packing disorder, but our best-fit value for b was consistent with published results [143]. Therefore, we used $b = 0.43$ for the remainder of this study. Hence, for external deformation, the solid–fluid transition point in the vertex model increases quadratically with tissue anisotropy Q.

We also tested how the model predictions change when we introduce anisotropy generated by internal forces into the vertex model. We modeled myosin planar polarity as increased tensions on "vertical" cell–cell contacts (Fig. 2.8b and Appendix A.2, Fig. A5) [127] and focused again on stationary, force-balanced states. We investigated simulations of model tissues with internal forces, both with (Fig. 2.8b) and without (Fig. 2.8b, Inset) externally applied deformation. We found that in both cases, solid states exist for larger cell shape indices than the isotropic $p_0^* = 3.94$, and our results were again consistent with the fit from Fig. 2.8a (solid lines in Fig. 2.8b and Fig. 2.8b, Inset). With finite anisotropic internal tensions only, we obtained states in the fluid regime that do not reach a force-balanced state (see detailed discussion in Appendix A), and this explains the white region devoid of stable states in the upper middle region of Fig. 2.8b, Inset. Taken together, these findings demonstrate that a combination of cell shape $\overline{p}$ and cell shape alignment

Q in the vertex model indicates whether an anisotropic tissue is in a solid-like or fluid-like state, regardless of the underlying origin of anisotropy.

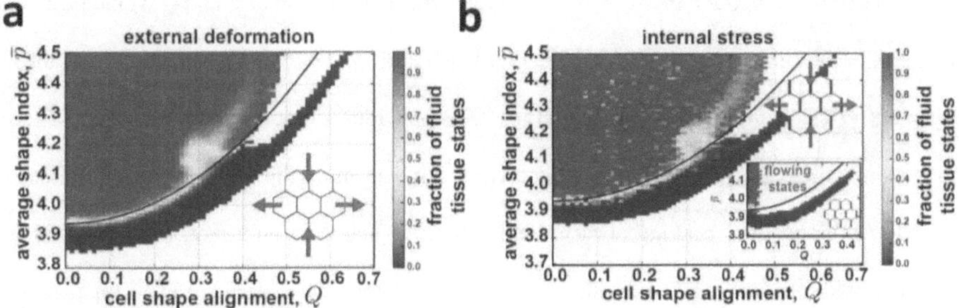

Figure 2.8. The solid-to-fluid transition in a vertex model of anisotropic tissues. Vertex model simulations for the case of an anisotropic tissue arising due to externally induced deformation (a) (cf. Fig. 2.7a and b), due to internal active stresses generated by an anisotropic cell–cell interfacial tension combined with externally applied deformation (b), and due to internal active stresses without any externally applied force (b, Inset) (Appendix A). The fraction of tissue configurations that are fluid is plotted as a function of $\bar{p}$ and Q. For both internal and external sources of anisotropy, the critical shape index $\bar{p}$ marking the transition between solid states (blue) and fluid states (red) is predicted to depend quadratically on Q. White regions denote combinations of $\bar{p}$ and Q for which we did not find force-balanced states. In particular, in the case of finite tension anisotropy, we did not find any stable force-balanced fluid states, and the red fluid states in b all correspond to the limiting value of zero-tension anisotropy. In Appendix A.1, Supplementery materials and methods, we explain how the lack of fluid states for finite tension anisotropy can be explained analytically. Our findings quite generally suggest that stationary states of fluid tissues with an anisotropic cell–cell interfacial tension are difficult to stabilize even when preventing overall oriented tissue flow via the boundaries. In a, the solid line shows a fit of the transition to Eq. 2.2 with $p_0^* = 3.94$ and $b = 0.43$; b and b, Inset show this same line. In a, a deviation from Eq. 2.2 is only seen around $\bar{p} \sim 4.15$ and $Q \sim 0.3$, where we observe an abundance of solid states, which is likely due to the occurrence of manyfold vertices in this regime (Appendix A.2, Fig. A3), which are known to rigidify vertex model tissue [140].

2.3.4 Cell shape and cell shape alignment together indicate the onset of cell rearrangement during *Drosophila* axis elongation

We returned to our experiments to test whether a combination of $\bar{p}$ and Q would be a better predictor for the behavior of the *Drosophila* germband during convergent extension. We quantified alignment Q using the triangle method (Fig. 2.9a) and found that, prior to the onset of tissue elongation, which begins at $t = 0$ min, alignment is not very high (Fig. 2.9b). Q began to increase just prior to elongation, peaking at $t = 1$ min (Fig. 2.9b and Appendix A.2, Fig. A6), which is consistent with observations using other cell-pattern metrics [79, 127, 131, 133]. This peak in Q corresponds to stretching of cells along the DV axis, perpendicular to the axis of germband extension, and coincides with the time period during which the presumptive mesoderm is invaginating [133, 144]. Q relaxes back to low levels during axis elongation (Fig. 2.9b). Plotting $\bar{p}$ versus Q at each time point from movies of individual wild-type embryos revealed common features, despite embryo-to-embryo variability (Fig. 2.9c, Inset). Initially, we saw a concomitant increase of $\bar{p}$ and Q prior to the onset of convergent extension. Above $\bar{p} = 3.87$, Q decreased drastically as $\bar{p}$ continued to increase, indicating that further increases in $\bar{p}$ are associated with randomly oriented cell shapes (cf. Fig. 2.7c). Thus, cell shapes in the germband are transiently aligned around the onset of convergent extension.

We next asked whether this temporary increase in alignment could help resolve the seeming contradiction between the measured cell shapes and cell rearrangement rates. To this end, we investigated how $\bar{p}$ and Q correlate with the instantaneous rate of cell rearrangements occurring within the germband, with higher rearrangement rates associated with more fluid-like behavior and/or larger active driving forces (Fig. 2.9c). The anisotropic vertex model predicts that the solid or fluid behavior of the tissue should depend on both $\bar{p}$ and Q according to Eq. 2.2, with only two adjustable parameters, p_0^* and b. We fit Eq. 2.2 to our experimental data by minimizing

a quality-of-fit measure defined as the number of experimental data points on the wrong side of

the theoretical transition line, and for simplicity varied only p_0^* while keeping the theoretically

determined value for b. Varying the value b leads, at most, to a slight improvement of our fit

(Appendix A.2, Fig. A7). To differentiate between solid-like and fluid-like tissue behavior in the

experimental data, we need to choose a cutoff value for the cell rearrangement rate. Choosing a

cutoff of 0.02 min−1 per cell yielded a best fit with $p_0^* = 3.83$ (solid line, Fig. 2.9c). To confirm

that our prediction of a quadratic dependence on Q is supported by the data, we also identified the

best fit to a null hypothesis of a Q-independent transition point (horizontal dashed line, Fig. 2.9c).

Using our quality-of-fit measure, we found that the Q-dependent fit was always better, independent

of the chosen cell rearrangement rate cutoff (Appendix A.2, Fig. A7).

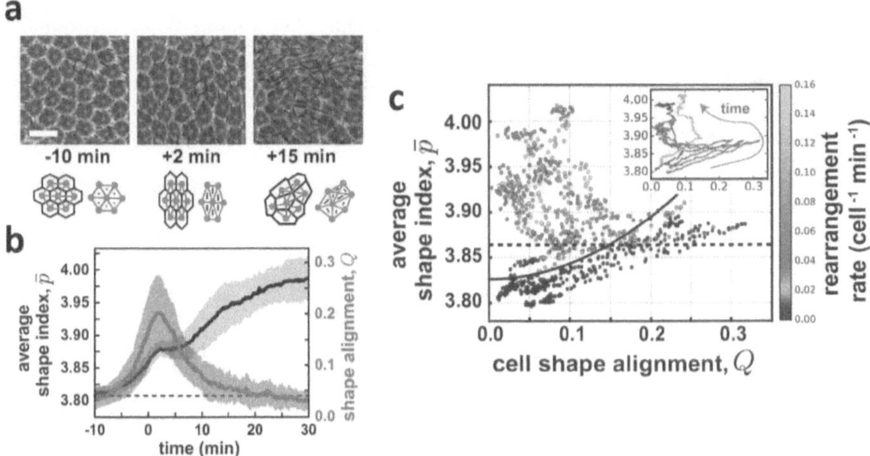

Figure 2.9. Cell shape and cell shape alignment together predict the onset of cell rearrangements during *Drosophila* convergent extension. (a) Confocal images from time-lapse movies of epithelial cell patterns in the ventrolateral region of the germband during *Drosophila* axis elongation. Cell outlines were visualized with gap43:mCherry [144]. Anterior left, ventral down. (Scale bar, 10 μm.) Images of cells with overlaid triangles that were used to quantify cell

shape anisotropy. Cell centers (green dots) are connected with each other by a triangular network (red bonds). Cell shape stretches are represented by triangle stretches (blue bars), and the average cell elongation, Q, is measured [147]. (b) The cell shape alignment index Q (red) and average cell shape index $\bar{p}$ (black, same as Fig. 2.5b) for the germband tissue before and during axis elongation. Q was calculated for each time point, and the mean and SD between embryos is plotted (n = 8 embryos with an average of 306 cells analyzed per embryo per time point). The onset of tissue elongation occurs at $t = 0$. The dashed line denotes the reported value for the solid–fluid transition in the isotropic vertex model, $\bar{p} = 3.811$ [67]. (c) The relationship between $\bar{p}$ and Q for eight individual wild-type embryos, with each point representing $\bar{p}$ and Q for a time point in a single embryo. Instantaneous cell rearrangement rate per cell in the tissue is represented by the color of each point, with blue indicating low rearrangement rates and red to yellow indicating high rearrangement rates. The black solid line indicates a fit to Eq. 2.2 with a rearrangement-rate cutoff of 0.02 min^{-1} per cell (Appendix A.1, Supplementary materials and methods), from which we extract $p_0^* = 3.83$, where b was fixed to the value obtained in vertex model simulations (cf. Fig. 2.8a). (c, Inset) $\bar{p}$ and Q for individual embryos over time.

Comparing the trajectories of individual embryos (Fig. 2.9c, Inset) to the predicted transition in the anisotropic vertex model (Fig. 2.9c), we see that, during early times, when $\bar{p}$ and Q are both increasing, the tissue stays within the predicted solid-like regime. The subsequent rapid decrease in Q brings embryos closer to the transition line. As $\bar{p}$ further increases, individual embryos cross this transition line, which coincides with increased rates of cell rearrangement, at different points $(Q, \bar{p})$. Thus, compared to the isotropic model, the anisotropic vertex model better describes the onset of rapid cell rearrangement and tissue flow during convergent extension with two metrics of cell patterns, $\bar{p}$ and Q, that are both easy to access experimentally.

2.3.5 Accounting for cell shape alignment and cell packing disorder allows for a parameter-free prediction of tissue behavior

While the above results confirm that tissue anisotropy must be taken into account to predict the onset of rapid cell rearrangement, the theoretical prediction in Fig. 2.9c still required a fit parameter p_0^*. Theoretical results suggest that this fit parameter, which is the isotropic transition point in the absence of anisotropic forces, should depend systematically on cell packing disorder quantified by vertex coordination [140] and fraction of pentagonal cells (Fig. 2.6a).

Therefore, we analyzed the $\bar{p}$ and Q data for each embryo individually, by fitting them to Eq. 2.2 with $b = 0.43$, where we again used p_0^* as the only fit parameter (Fig. 2.10a, Inset). We compared the p_0^* obtained for each embryo (purple point, Fig. 2.10a, Inset) to the average vertex coordination number in the tissue at the time of the transition (green point, Fig. 2.10a, Inset) and found a clear correlation (dashed line, Fig. 2.10c), which fit well with the previous theoretical prediction [140], with no fit parameters.

Combining this previous theoretical prediction of the effects of vertex coordination on the solid–fluid transition in isotropic tissues with our prediction for how cell shape alignment shifts this transition in anisotropic tissues in Eq. 2.2 generates the following parameter-free prediction of the critical shape index for tissue fluidity:

$$\bar{p}_{crit} = 3.818 + (z - 3)/B + 4bQ^2 \tag{2.3}$$

where z is the measured average vertex coordination number, and the other parameters are universally determined a priori from vertex model simulations: $B = 3.85$ [140], and $b = 0.43$. To test this prediction, we plot the cell shape index corrected by the vertex coordination number, $\bar{p}_{corr} = \bar{p} - (z - 3)/B$, versus cell shape alignment Q in the germband of wild-type embryos and compared it to the theoretical curve given by $\bar{p}_{corr} = 3.818 + 4bQ^2$ (solid line, Fig. 2.10a). Remarkably, this parameter-free prediction described our experimental data well. We compared the quality of fit to alternative parameter-free predictions and found that Eq. 2.3 consistently provided the best prediction for a wide range of cell rearrangement rate cutoffs (Appendix A.2, Fig. A7).

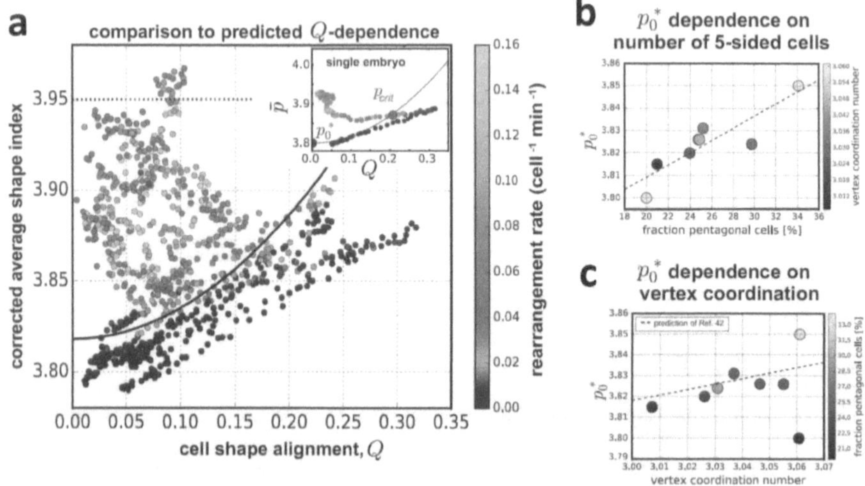

Figure 2.10. A parameter-free prediction of tissue mechanical behaviors by cell shape, cell shape alignment, and cell packing disorder. (a) The relationship between the corrected average cell shape index $\bar{p}_{corr}$ and cell shape alignment Q for eight individual wild-type embryos, with each point representing a time point in a single embryo. The cell shape index is corrected by the vertex coordination number z as $\bar{p}_{corr} = \bar{p} - (z - 3)/B$, with $B = 3.85$ [140]. Instantaneous cell rearrangement rate per cell in the tissue is represented by the color of each point. The solid line indicates the parameter-free prediction of Eq. 2.3. (a, Inset) Single embryo fit to Eq. 2.2. (b) p_0^* from single embryo fits to Eq. 2.2 correlate with the fraction of pentagonal cells f_5, a metric for cell packing disorder in the tissue, at the transition point. The dashed line represents a linear fit to the data. When using a rearrangement-rate cutoff of 0.02 min^{-1} per cell for the single embryo fits, we obtain for this linear fit $p_0^* = 3.755 + 0.27 f_5$. (c) p_0^* from single-embryo fits to Eq. 2.2 correlate with the average vertex coordination number, another metric for packing disorder in the tissue, at the transition point. The dashed line represents the previous theoretical prediction for how manyfold vertices influence tissue behavior [140].

Some embryos deviated from the theoretical prediction from ref. [140] (Fig. 2.10c), suggesting that perhaps alternate features of packing disorder may play an important role in those embryos. Thus, we also compared p_0^* obtained from the individual-embryo fits to the respective fraction of pentagons at the time of the transition and found a strikingly clear correlation well described by a linear relation (Fig. 2.10b, dashed line is a linear fit). This relationship quantitatively differs from what we extracted from our vertex model simulations (Fig. 2.6a), indicating again that other aspects of packing disorder may also play a role. Nevertheless, using this linear fit to correct the shape index for each data point by the fraction of pentagonal cells, we obtained an improved prediction of our data (compare Fig. 2.10a to Appendix A.2, Fig. A8) at the expense of requiring two fit parameters.

Taken together, these results show that we can quantitatively predict the behavior of the germband tissue in wild-type embryos, with no fit parameters using Eq. 2.3, from an image of cell patterns in the tissue. To do so, we needed to quantify three observables: cell shapes, cell alignment, and cell packing disorder. We found that vertex coordination and the fraction of pentagonal cells are both good proxies for packing disorder, in vertex model simulations and the germband.

2.3.6 Cell shape, alignment, and tissue behavior in *snail twist* and *bcd nos tsl* mutant embryos

Since the *Drosophila* germband experiences both internal forces due to myosin planar polarity and external forces from neighboring tissues, we wondered whether our theoretical predictions hold when altering the nature of the forces in the germband. To dissect the effects of internal and external sources of tissue anisotropy, we studied cell patterns in *snail twist* mutant embryos, which lack genes required for invagination of the presumptive mesoderm [148], and in *bcd nos tsl* (*bnt*) mutant embryos, which lack patterning genes required for planar-polarized patterns of myosin localization and axis elongation [41, 103].

First, we analyzed cell shapes and cell shape alignment in the germband of *snail twist* mutant embryos in which the presumptive mesoderm does not invaginate. In *snail twist* embryos, we observed that the germband tissue elongated (Fig. 2.11c) and cell rearrangements occurred (Fig. 2.11d), similar to prior studies [79], although at somewhat reduced rates compared to in wild-type embryos. However, in contrast to wild-type embryos, we found that the cell shape alignment Q was significantly reduced between $t = -5$ min and $t = +8$ min (Fig. 2.11a, b and f), similar to previous reports of other metrics for cell stretching [79]. The cell shape index $\bar{p}$ was also reduced during this period (Fig. 2.11a, b and e). These observations are consistent with the idea that external forces from mesoderm invagination produce the transient cell shape elongation and alignment observed in wild-type embryos.

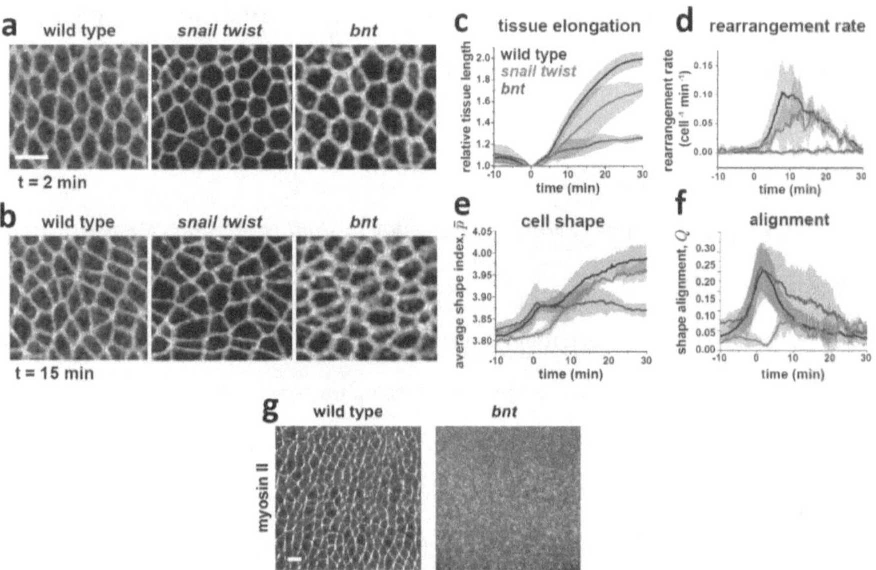

Figure 2.11. Cell shape, cell shape alignment, and cell rearrangement rates in the germband of *snail twist* and *bcd nos tsl* (*bnt*) mutant embryos. *snail twist* embryos lack ventral patterning genes required for presumptive mesoderm invagination. *bnt* embryos lack AP patterning genes required for axis elongation and show severely disrupted myosin planar polarity compared to wild type. (a and b) Confocal images from time-lapse movies of cell patterns at $t = +2$ min and $t = +15$ min. Cell outlines visualized with fluorescently tagged cell membrane markers: gap43:mCherry in wild type, Spider:GFP in *snail twist*, and Resille:GFP in *bnt*. Polygon

representations of cell shapes are overlaid (green). (Scale bar, 10 μm.) (c) Tissue elongation is moderately reduced in *snail twist* and severely reduced in *bnt* compared to wild type. (d) Cell rearrangement rate is moderately decreased in *snail twist* and severely reduced in *bnt*. (e) In *snail twist*, the average cell shape index $\bar{p}$ is reduced compared to in wild type for -5 min $< t < 5$ min. In *bnt*, $\bar{p}$ shows similar behavior to in wild type for $t < 5$ min, but does not show further increases with time for $t > 5$ min. (f) In *snail twist*, the cell alignment index Q is strongly reduced for -5 min $< t < 10$ min compared to in wild type. In *bnt*, Q shows similar behavior to in wild type for $t < 5$ min, but relaxes more slowly to low levels. (c-f) The mean and SD between embryos is plotted (three *snail twist* and five *bnt* embryos with an average of 190 cells per embryo per time point). (g) *bnt* mutant embryos, which lack anterior-posterior patterning genes required for axis elongation, show severely disrupted myosin planar polarity compared to wild-type embryos. Scale bar, 10 μm.

Next, we tested whether our theoretical predictions would describe tissue behavior in *snail twist* embryos, even with their significantly reduced cell alignment. We found that the onset of rapid cell rearrangement in *snail twist* embryos was also well predicted by Eq. 2.3 (Fig. 2.12a). This was corroborated by comparing the parameters p_0^* of the individual *snail twist* embryo fits to the vertex coordination number at the transition (Fig. 2.12a, Inset), which is close to the previous theoretical prediction (dashed line) [140]. Hence, our prediction also held in embryos with reduced cell shape alignment Q, where the transition to rapid cell rearrangement occurred at a lower cell shape index $\bar{p}$ compared to in wild-type embryos (Fig. 2.12b).

To investigate how disrupting other forces in the germband affects tissue behavior, we studied cell patterns in *bnt* mutant embryos, which lack AP patterning genes required for axis elongation. These mutant embryos did not display myosin planar polarity, although there was significant myosin present at the apical cortex of cells (Fig. 2.11g). The *bnt* embryos had severe defects in tissue elongation (Fig. 2.11c), cell rearrangement (Fig. 2.11d), and endoderm invagination, but still underwent mesoderm invagination [40, 41, 64, 79, 103, 131]. $\bar{p}$ displayed an initial increase (Fig. 2.11e), concomitant with an increase in Q (Fig. 2.11f), similar to in wild-type embryos. After $t = 1$ min, $\bar{p}$ did not increase further and took on a steady value of 3.87 (Fig. 2.11e). This supports the idea that the further increase in $\bar{p}$ in wild-type embryos is due to internal anisotropies associated with myosin planar polarity or external forces associated with endoderm invagination. Interestingly, Q returned more slowly to low levels in *bnt* compared to wild-type embryos (Fig. 2.11f), suggesting a potential role for myosin planar polarity, cell rearrangements oriented along the AP axis, or endoderm invagination in relaxing cell shape alignment along the DV axis. The *bnt* tissues did not transition to a state of rapid cell rearrangement. This was not consistent with the predictions of Eq. 2.3 (Fig. 2.12c), which predicts some fluid-like tissue states in the germband of *bnt* embryos, suggesting that either the driving forces are too small or that there are additional barriers that prevent rapid cell rearrangement in these embryos.

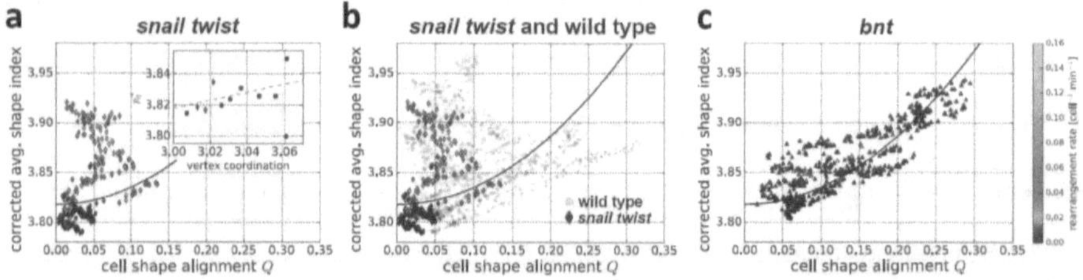

Figure 2.12. Tissue mechanical behavior prediction in the germband of *snail twist* and *bcd nos tsl (bnt)* mutant embryos. Relationship between the corrected cell shape index $\bar{p}_{corr}$ and Q for three *snail twist* (a and b), eight wild-type (b), and five *bnt* (c) embryos, with each point representing a time point in a single embryo. Instantaneous rearrangement rate is represented by the color of each point. Solid lines represent the prediction of Eq. 2.3. (b) Tissue behavior in *snail twist* and wild-type embryos, all of which exhibit rapid cell rearrangement during convergent extension, is well described by the prediction of Eq. 2.3, which does not require any fitting parameters. Avg., average.

Taken together, these findings demonstrate that external forces associated with mesoderm invagination contribute to tissue anisotropy in the germband and that the onset of rapid cell rearrangement can be predicted from cell shape and alignment, even in the absence of forces associated with mesoderm invagination.

2.4 Discussion

In this work, we show that cell shape, cell alignment, and packing disorder can be used to understand and predict whether an anisotropic tissue flows and remodels like a fluid or, instead, maintains its shape like a solid. Importantly, in contrast to isotropic tissues, the mechanical behavior of the converging and extending *Drosophila* germband cannot be predicted by cell shape and packing disorder alone. Instead, we show via theoretical analysis and simulation that, in anisotropic tissues, three experimentally accessible metrics—the cell shape index $\bar{p}$, the cell alignment index Q, and packing disorder quantified by either vertex coordination or fraction of pentagonal cells—are required to determine whether an anisotropic tissue flows and remodels or not. We demonstrate that the onset of rapid cell rearrangement in wild-type *Drosophila* embryos is indeed more accurately described by a combination of these three cell-pattern metrics, using an equation with no fit parameters, than by cell shape or packing disorder alone. We further tested this prediction in *snail twist* mutant embryos in which the presumptive mesoderm does not invaginate and found that our parameter-free prediction successfully predicted the onset of rapid cell rearrangement and tissue flow in this case as well. These findings suggest that convergent extension of the *Drosophila* germband might be viewed as a transition to more fluid-like behavior to help accommodate dramatic tissue flows. This raises the possibility that the properties of developing tissues might be tuned to become more fluid-like during rapid morphogenetic events.

A fluid-to-solid jamming transition has recently been reported in mesodermal tissues during zebrafish body axis elongation [105]. In contrast to the zebrafish mesoderm in which the transition to more solid-like behavior is associated with an increase in cellular volume fraction (proportion of the tissue occupied by cells), the *Drosophila* germband epithelium comprises tightly packed cells, and its mechanical behavior changes in the absence of any change in cellular volume fraction. Future studies will be needed to explore how the properties of epithelial cells might be regulated during development to tune the mechanical behaviors of the tissues in which they reside.

The vertex model predictions of tissue behavior are independent of the underlying origin of anisotropy, and therefore can be used to predict mechanical behavior of tissues from cell shape patterns, even when external and internal stresses cannot be directly measured. Although our current simulations were not able to access some of the tissue states driven by internal stresses, we found that the cases that were accessible were fully consistent with our simulation results without internal stresses. Importantly, the average cell shape index $\bar{p}$, cell shape alignment index Q, and metrics for packing disorder are easy to access experimentally from snapshots of cell packings in tissues, even in systems where time-lapse live imaging of cell rearrangement and tissue flow is not possible. Thus, this approach may prove useful for studying complex tissue behaviors in a broad range of morphogenetic processes occurring in developing embryos *in vivo* or organoid systems *in vitro*.

In our analysis, we characterized the mechanical state of the germband epithelial tissue using the rate of cell rearrangement as the observable. We made this choice because direct measurements of the mechanical properties of the germband remain a significant experimental challenge [45, 117, 120]. Generally, higher rates of cell rearrangement could be due to more fluid tissue properties or a stronger driving force, which is the sum of externally applied forces and internally generated mechanical stresses. Based on our Eq. 2.3 result, the cell shape index and alignment predict the onset of rapid cell rearrangement in the germband. While this would be consistent with the tissue becoming more fluid, it is also possible that the observed increase in cell rearrangement rate is, at least in part, due to an increase in the driving force while the tissue remains solid.

To parse this possibility further, it is useful to consider a solid tissue, where the tissue will flow only if it is pulled with a force above some threshold called the yield stress. If the tissue is deeply in the solid state, far from the solid–fluid transition, and the applied force is far above the yield stress, one would expect cells to acquire elongated shapes and transiently form manyfold vertices during cell rearrangements in response to the applied force. The rearrangement rate would correlate with the cell shape index, after accounting for packing disorder and alignment, which is similar to what we predict with our fluid–solid model. However, based on our vertex model simulations, we would not expect to see tissue states with high shape index $\bar{p}$ and low alignment Q associated with high rearrangement rates for solid tissues. Since we do observe such tissue behavior during germband extension, this suggests that the germband is more fluid-like during these periods with high cell rearrangement rates.

Of course, it could be that the tissue is a very weak yield-stress solid, so that it becomes fluid-like under very small applied forces. This is consistent with the observations that the large majority of rearrangements are oriented along the head-to-tail body axis [41, 42, 103, 104, 126], and the time period of rapid cell rearrangement (Fig. 2.4b) coincides with the period of planar-polarized myosin [64, 104, 129]. Direct mechanical measurements of the germband have not been conducted during axis elongation, but ferrofluid droplet and magnetic-bead microrheology measurements have probed the mechanical behavior of the epithelium prior to germband extension in the cellularizing embryo. These studies report that tissue behavior is predominantly elastic (solid-like) over timescales less than several minutes and suggest fluid-like behavior on the longer, ~30-min timescales relevant for germband extension [51, 56]. These measurements might also be consistent with a weak yield-stress solid, an interpretation that would be supported by the near absence of cell rearrangements prior to germband extension. Taken together, these observations suggest that, over the time period that we describe the germband as "fluid-like," it could actually be a very weak yield-stress solid.

Though there is often little functional difference between a fluid and weak yield-stress solid, the difference may be relevant for mutant *bnt* embryos, whose behavior is not well captured by our theoretical predictions. In particular, we observed *bnt* tissues with $\bar{p}$, Q, and cell packing disorder that would be predicted to display fluid-like behavior, but did not undergo rapid cell rearrangement. This suggests that in these embryos, the driving forces are not sufficient to overcome the yield stress. One obvious explanation for this is that the germband in *bnt* embryos experiences altered forces associated with disrupted myosin planar polarity [103] and defects in endoderm invagination, which would contribute to a reduced driving force. Alternatively, additional barriers to cell rearrangement in *bnt* mutants, of the sort described in ref. [154], could also explain this behavior.

Similarly, our vertex model does not predict the observed decrease in cell rearrangement rates after 20 min of axis elongation (Fig. 2.4b). Given the observed high values of $\bar{p}$ and low values of Q, our model would still predict fluid-like behavior. Just as in the *bnt* mutants, this discrepancy could be explained by a decreased driving force or additional barriers to cell rearrangement. The former explanation is supported by the observation that myosin planar polarity reaches a maximum 5 to 10 min after the onset of axis elongation and then decreases during the rest of the process [79, 104, 129], while the latter could potentially be explained by maturation of cell junctions or changes to adhesive interactions over the course of embryonic development [155, 156].

Consistent with the notion of additional barriers to cell rearrangement, recent work suggests that local remodeling of active junctional tension at cell–cell contacts only occurs above a critical strain threshold in cultured epithelial cells [154, 157]. This is consistent with a growing body of work that points toward important roles for membrane trafficking and E-cadherin turnover in junctional remodeling during *Drosophila* epithelial morphogenesis [53, 158-160]. Indeed, such a mechanism of mechanosensitive barriers to junctional remodeling and cell rearrangement can be added to standard vertex models to explain such weak yield-stress behavior [154].

Moving forward, it will be interesting to explore experimentally how the nature of internal and external forces contribute to tissue mechanics, cell rearrangement, and tissue flows in the germband and other developing epithelial tissues. Incorporating these features into more sophisticated vertex models will contribute to understanding the diverse behaviors of living tissues, and the approaches we develop here will be useful for interrogating these questions.

In the next chapter, we study one of the key cellular machineries in epithelial tissues that is likely to influence tissue mechanical behavior and test how perturbing this machinery influences tissue mechanical behaviors. In particular, we investigated the role of E-cadherin-based cell-cell adhesion in epithelial tissue mechanics.

Chapter 3. Molecular Mechanisms Underlying Tissue Mechanical Behavior --- The Role of Cell-Cell Adhesion in Epithelial Tissue Mechanics

In Chapter 2, we explore how epithelial tissues behave as remarkable materials whose mechanical properties are tuned for specific morphogenetic events [72]. We show that epithelial tissue mechanical properties can be predicted by cell shape, cell alignment, and cell packings within the tissue [72]. However, how the properties of epithelial cells are regulated to tune the mechanics of the tissues in which they reside, remains unknown. Contractile tension and cell-cell adhesion are thought to be key machineries that control tissue mechanics. Cell-cell adhesion physically adheres cells to each other to make the tissue more solid-like, but also organizes and transmits forces that promote tissue remodeling to make it more fluid-like. It is unclear how cell-cell adhesion enacts its dual functions in epithelial tissue mechanics and morphogenesis. To explore the role of cell-cell adhesion in controlling epithelial tissue mechanics, in Chapter 3, genetic approaches are used to tune cell-cell adhesion levels in the *Drosophila* embryo, and live imaging and quantitative image analysis are combined to study the effects of cell-cell adhesion levels on cellular and tissue behaviors. Our results show that cell shape, cell rearrangement, and tissue fluidity all display biphasic dependencies on cell-cell adhesion levels and these dependencies are linked with the coupling of cell-cell adhesion to the cellular actomyosin cytoskeleton. This work advances our fundamental understanding of how molecular-level activities control tissue-level mechanical behaviors, and provides quantitative, multi-scale, experimental approaches to explore tissue mechanics *in vivo*. This is essential to revealing mechanical mechanisms underlying epithelial tissue morphogenesis and sheds light on the improper regulation of tissue mechanics associated with birth defects, aberrant wound healing, and cancer metastasis. These experimental results provide the necessary input for building and testing models of epithelial tissue mechanics and may be useful for building tissues with precise shapes, structures, and mechanical properties.

3.1 Introduction

3.1.1 Epithelial tissue morphogenesis

To sculpt the diverse and elaborate biological structures in multicellular organisms, tissues undergo a wide variety of morphogenetic processes. Epithelia form sheets of cells organized in monolayers, as in developing embryos and the gut, or in multilayered as in the skin [175]. Epithelial sheets play vital roles in physically supporting the structure of embryos and organs, serving as effective barriers against pathogens, and chemically separating different physiological environments [3, 4, 85, 113, 175]. These vital functions require robust and cohesive tissue structures formed by tightly associated epithelial cells (Fig.1.7a). Remarkably, even when cells are tightly packed to maintain tissue integrity, epithelial tissues are also dynamic and can display dramatic reorganization and remodeling, such as during convergent extension movements that narrow and elongate tissues (Fig. 1.7b, Fig. 2.1a, and Fig. 3.1a); convergent extension movements are highly conserved among animals and used in elongating tissues, tubular organs, and overall body shapes [86, 87, 122]. During *Drosophila* embryonic development, for example, the germband epithelium rapidly converges and extends, doubling the length of the body axis in just 30 minutes while maintaining tissue cohesion (Fig. 1.7b, Fig. 2.1a, and Fig. 3.1a) [41, 45]. This convergent extension movement in the *Drosophila* embryo is often referred to as germband extension (GBE) as it involves elongation of the germband epithelium. How tissues can remodel and flow while maintaining their overall mechanical integrity remains poorly understood.

3.1.2 Mechanics in epithelial tissue morphogenesis

Morphogenesis is fundamentally mechanical in nature [43-47]. Mechanical properties that determine how tissues resist deformation by mechanical forces, for example, are likely to play key roles in regulating and controlling morphogenesis. Recently, experimental studies in both plants and animals have highlighted that epithelial tissue remodeling can be linked with tissue fluidity [18, 53, 72, 105, 118, 176]. Fluid-like tissues accommodate tissue flow and remodeling, while solid-like tissues resist flow. Meanwhile, computational models have highlighted physical mechanisms by which epithelial tissues might transition between solid-like and fluid-like behaviors [64, 67, 119]. In these models, there are typically higher energy barriers for cells to rearrange in solid-like tissues, while lower energy barriers for cells to rearrange are seen in fluid-like tissues [67, 69, 70]. Our work in Chapter 2 analyzing the *Drosophila* germband epithelium and using anisotropic vertex model simulations demonstrate that mechanical properties of tissues appear to be tuned to become more fluid-like during rapid morphogenetic events. We also propose that tissue fluidity in anisotropic epithelial tissues can be predicted by two experimentally accessible metrics of cell patterns – cell shape and cell alignment (Fig. 2.8 and Fig. 2.10). These prior studies, however, did not address how tissue fluidity might be tuned by the properties of epithelial cells and their molecular composition.

In addition, morphogenesis requires either external forces that deform the tissue or asymmetries in cell behaviors that internally drive tissue-shape change, both of which are often associated with the asymmetric localization of key molecules inside cells [122-125]. In epithelial tissues, cells can change shape or rearrange as tissues deform during morphogenesis (Fig. 1.7c and Fig. 2.4a). From a mechanical point of view, the spatiotemporal deployment and coordination of contractile tension and cell-cell adhesion (Fig. 1.7d and Fig. 3.1b) are thought to be key mechanisms controlling these epithelial cellular and tissue mechanical behaviors such as cell shape change, rearrangement, tissue extension, folding, and invagination [43-45, 63]. For instance, the extent of cell contacts is determined by adhesion forces that stabilize cell-cell interfaces, balanced against cortical tension exerted by the actomyosin network that tends to reduce cell contact lengths [91]. In the *Drosophila* embryo, epithelial germband tissue extension is largely driven by oriented local cell rearrangements, accompanied by cell shape changes (Fig. 2.4) [41, 79, 103, 126]. *Drosophila* germband extension requires the planar-polarized localization of the force-generating motor protein myosin II, which is specifically enriched at cell edges that are oriented perpendicular to the head-to-tail body axis [40, 103, 104, 126-129], and external forces from neighboring tissues including the mesoderm and endoderm, which influence cell shapes in the germband during convergent extension [131, 132] (Fig. 2.1). Despite being fundamental to epithelial tissue behavior *in vivo*, it is unclear how the balance between cell-cell adhesion and contractile tension controls or influences epithelial cell and tissue mechanical behaviors.

Computational models, such as vertex, particle jamming, cellular Potts, and continuum models, have been used to simulate epithelial tissue mechanics [58-62]. Contractile tension and cell-cell adhesion are key parameters in many of these models. For example, recent vertex models predict that epithelial tissues can become more solid-like or more fluid-like depending on the balance between cell tension and cell-cell adhesion [67, 69, 70]. However, few *in vivo* experimental studies have been done to inform these models and test their predictions. In a simple, intuitive model, for example, increased cell-cell adhesion is predicted to result in slower cell rearrangement associated with additional time required to disassemble and remodel cell junctions as cells exchange neighbors. In an alternate model, increased cell-cell adhesion is predicted to result in faster cell rearrangement due to longer cell-cell contacts, more elongated cell shapes, and likely lower energy barriers for cell rearrangement [67]. *In vivo* quantitative experiments of how cell-cell adhesion affects epithelial tissue mechanics are required to distinguish between these models. Another issue is the challenge of connecting parameters in models to the actual molecular and physical factors in biological systems. For example, although the vertex model succeeded in predicting epithelial tissue fluidity in some contexts, the molecular contributions to the model parameters remain unclear [67, 72]. Therefore, systematic, quantitative experimental studies on epithelial tissue mechanics are required to build more precise and predictive models of epithelial tissues.

3.1.3 Dual roles of cell-cell adhesion in epithelial tissue mechanics

At the molecular level, epithelial cohesion requires E-cadherin molecules to form adhesive contacts within the tissue. E-cadherin comprises extracellular, transmembrane, and intracellular domains to connect neighboring cells, which are further stabilized by binding to actin filaments via catenin molecules such as β-catenin and α-catenin, vinculin, and other proteins [96, 175] (Fig. 3.1). E-cadherin clusters often concentrate at the level of adherens junctions and form an adhesive belt such as the *zonula adherens* that stitches cells together [96, 175]. Therefore, E-cadherin adhesions reflect both the ligation of E-cadherin ectodomains [183, 184] and the mechanical coupling to the actomyosin network via α-catenin and vinculin [185-189], which stabilizes E-cadherin complexes. As a consequence, adhesion complexes are under tension and transmit subcellular forces exerted by actomyosin networks to the cortex [188, 189] (Fig. 3.1b).

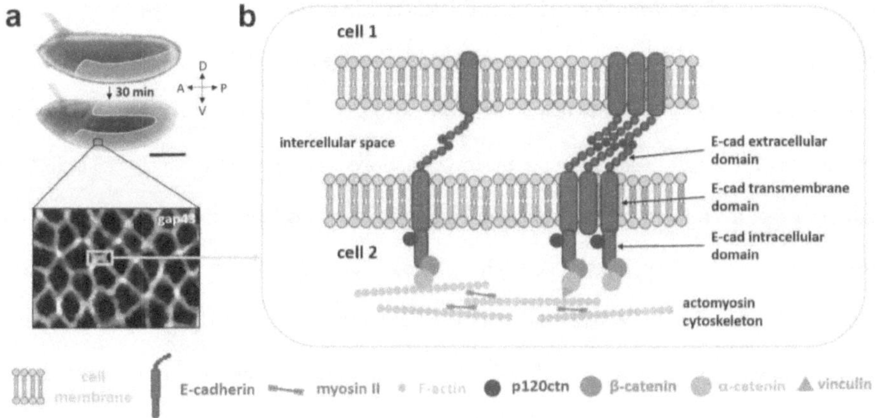

Figure 3.1. Molecular mechanisms underlying *Drosophila* body axis elongation. (a) Top: Images from bright-field movies before and during *Drosophila* body axis elongation. Anterior (A) to the left. Posterior (P) to the right. Dorsal (D) to the top. Ventral (V) to the bottom. Germband epithelial tissue, marked by yellow, narrows in DV direction and elongates in AP direction to double the length of body axis in just 30 min. Scale bar, 100 μm. Bottom: Images of *Drosophila* germband epithelium from confocal time-lapse movies during body axis elongation. Fluorescently labeled cell membranes. Cells take polygonal shapes and are tightly packed. (b) Schematic of molecules at cell-cell contacts showing the E-cadherin (dark blue) interface mediating cell-cell adhesion. The extracellular domains of E-cadherin from adjacent cells bind to each other by calcium activated dimerization. The intracellular domain binds to p120ctn and to β-catenin. The complex formed with β-catenin allows α-catenin or vinculin to link this complex to the cellular actomyosin cytoskeleton (red, orange). E-cadherin molecules can form clusters (right).

The role of cell-cell adhesion in epithelial tissue mechanics is not well understood yet. Cell-cell adhesion seems to play interesting "dual" roles in epithelial tissue mechanics and morphogenesis by both inhibiting and promoting tissue remodeling. On the one hand, cell-cell adhesion physically connects cells to maintain tissue cohesion, which is predicted to inhibit tissue flow. On the other hand, it also organizes and transmits forces and tensions that promote tissue flow. Therefore, it is uncertain how cell-cell adhesion enacts its dual functions to impact epithelial tissue mechanics and behaviors.

Cell-cell adhesion molecules such as E-cadherin are coupled with actomyosin cytoskeleton (Fig. 3.1b) and are regulated by multiple mechanisms such as mechanical loading, diffusion and endocytosis [183-192]. Not only can actomyosin stabilize E-cadherin, remarkably, recent work reveals that E-cad can also influence actomyosin flows and contractile tensions [159, 193], indicating that the role of cell-cell adhesion in morphogenesis is likely complex. Therefore, it is essential to understand the roles of cell-cell adhesion in epithelia to have a more comprehensive understanding of the molecular and mechanical mechanisms underlying epithelial tissue morphogenesis. The effects of cell-cell adhesion on tissue mechanics *in vivo* are poorly characterized, in part due to the lack of sophisticated methods to measure mechanics inside embryos and the strong coupling of mechanical and biological factors in tissues, which makes it difficult to dissect mechanisms.

3.1.4 Systematic, quantitative, experimental studies to explore epithelial tissue mechanics *in vivo*

To explore the role of cell-cell adhesion in epithelial tissue morphogenesis *in vivo*, dissect how cell-cell adhesion tunes epithelial tissue mechanics, and build more accurate models of epithelial tissues, in Chapter 3, I systematically modulated cell-cell adhesion levels in the epithelial tissue and studied the effects on tissue remodeling *in vivo*. I used molecular genetics approaches to modify E-cadherin expression in the *Drosophila* embryo, and combined live confocal imaging and quantitative image analysis to study the effects of E-cadherin levels on cellular behaviors and tissue mechanics during *Drosophila* body axis elongation. Our results show biphasic dependencies of cell shape, tissue fluidity, and cell rearrangement speed on E-cadherin levels, challenging current ideas of how adhesion influences tissue behaviors. These results reveal surprising links between cell adhesion, cell rearrangement, and tissue fluidity. In particular, tissues comprising cells with both lower or higher E-cadherin levels compared to wild-type controls tend to rearrange faster and behave in a more fluid-like manner. By studying the behaviors of molecules that interact directly or indirectly with E-cadherin, we find that E-cadherin levels tune the localization and dynamics of other molecules, including junctional myosin II, to influence cellular behaviors and tissue mechanics.

These findings suggest that E-cadherin-mediated cell-cell adhesion plays an active and nuanced role in modulating epithelial tissue mechanics and remodeling through both direct and indirect effects, which elucidates how tissues remodel while maintaining integrity and sheds light on how the improper regulation of adhesion might contribute to defects in tissue mechanics associated with birth defects, aberrant wound healing, and cancer metastasis. These experimental results serve as suitable inputs for building and testing models of epithelial tissue mechanics, including molecular contributions to tissue mechanics and regulation of cellular rearrangement time scales. This work also provides a foundation for strategies to control tissue fluidity and a wide variety of applications in engineering and medicine.

3.2 Materials and methods

In this study, *Drosophila* culture and genetics, confocal microscopy, quantitative image analysis, fluorescence recovery after photobleaching (FRAP), and immunohistochemistry were used to explore the role of cell-cell adhesion in epithelial tissue mechanics and morphogenesis.

Embryos were imaged and FRAP was conducted on a Zeiss LSM880 laser-scanning confocal microscope. E-cadherin levels were decreased by inhibiting expression using RNAi or increased by overexpression using a shg:GFP transgene. Protein intensities were quantified and analyzed by ImageJ [149], EPySeg [84] and custom codes. For live confocal imaging, the germband epithelial tissue is from *Drosophila* embryos generated at either 18°C or 23°C and analyzed *in vivo* at room temperature. Cell outlines were visualized with gap43:mCherry cell-membrane markers [144], E-cadherin molecules were visualized with GFP fluorescent markers, and myosin II and moesin molecules were visualized with mCherry fluorescent markers. Time-lapse movies were analyzed with SEGGA software in MATLAB [79] for topologies and dynamics, PIVlab (Version 1.41) in MATLAB [145] for quantifying tissue elongation, and custom code for quantifying cell alignment using the triangle method [130, 146, 147]. For fixed sample imaging, the germband epithelial tissue is from *Drosophila* embryos generated at either 18°C or 23°C and fixed at room temperature. E-cadherin molecules were visualized with Alexa-488 fluorescent markers and F-actin molecules were visualized by Alexa-647-conjugated phalloidin. Images from fixed samples were analyzed with ImageJ [149] for image processing and manual measurements of protein intensities, EPySeg [84] for automated image segmentation, and custom code for automatic measurements of protein intensities.

Unless otherwise noted, error bars are the standard deviation (SD). Mean values from normal distributions were compared by one-way ANOVA and the Tukey HSD *post-hoc* tests.

Methods for time-lapse imaging, tissue elongation measurement, automated image segmentation and cell rearrangement analysis, and cell shape index p and cell shape alignment Q analysis are already described in Chapter 2, Section 2.2 Materials and methods and Appendix A.1 Supplementary materials and methods.

The data that support the findings of this study are included in Chapter 3 and Appendix B in this book. Details of the materials and methods are listed below and protocols used in this study can be found in Appendix B.1.

3.2.1 Fly stocks and genetics

Fly stocks for systematic manipulation of E-cadherin levels. Embryos were generated at 18°C or 23°C. The DE-cadherin gene *shotgun* is abbreviated as *shg*. E-cadherin levels were modulated by transgenic overexpression or RNAi using the Gal4/UAS system. Maternal-tubulin-Gal4 drivers mat67 and mat15 activate genes encoded downstream of the UASp sequences used in this study (FlyBase FBto0000343). The activation level increases when temperature decreases for these drivers. To decrease E-cadherin levels, *shg* expression was partially inhibited by RNAi, using a UASp promoted TRiP transgene (Transgenic RNAi Program [194]). To increase E-cadherin levels, the UASp-shg:GFP transgene (a fusion between *shg* and GFP under the UASp promoter) was integrated into landing site attP2 on the third chromosome. The control group has the endogenous *shg* replaced with shg:GFP (a fusion between *shg* and GFP under the endogenous promoter on the second chromosome). In addition, Oregon R was used as a wild-type control (Appendix B, Fig B3a and b).

To visualize cell membranes, one maternal copy of a sqh-gap43:mCherry transgene (a fusion between mCherry and membrane-targeting sequences T:Rnor\Gap43 ubiquitously expressed under the sqh promoter) was integrated into landing sites attP40 or attP2, on the second or third chromosome, respectively. Fly stocks of the following genotypes are in ascending order of predicted E-cadherin levels.

> sqh-gap43:mCherry→attP40/+; TRiP/mat15 (18°C)
>
> sqh-gap43:mCherry→attP40/+; TRiP/mat15 (23°C)
>
> shg:GFP/mat67; sqh-gap43:mCherry→attP2/mat15 (23°C)
>
> mat67/mat67; UASp-shg:GFP→attP2/ UASp-shg:GFP→attP2 (23°C)
>
> sqh-gap43:mCherry→attP40/mat67; UASp-shg:GFP→attP2/mat15 (23°C)

sqh-gap43:mCherry→attP40/mat67; UASp-shg:GFP→attP2/mat15 (18°C)

Fly stocks for live imaging of myosin II and moesin. Embryos were generated at 23°C. The myosin II light chain gene *spaghetti squash* and the F-actin binding protein gene *Moesin* are abbreviated as *sqh* and *Moe*, respectively. E-cadherin levels were modulated as described above. To visualize myosin II, one maternal copy of a sqh-sqh:mCherry transgene (a fusion between mCherry and myosin regulatory light chain ubiquitously expressed under the sqh promoter) was integrated into landing sites attP2 on the third chromosome. To visualize actin filaments, one maternal copy of a sqh-Moe:mCherry transgene (a fusion between mCherry and the actin-binding ERM protein Moesin ubiquitously expressed under a sqh promoter) to label cell membranes was integrated into landing sites attP2 on the third chromosome. Embryos from the following genotypes were collected and imaged.

shg:GFP/+; sqh-sqh:mCherry, mat15/TRiP

shg:GFP/mat67; sqh-sqh:mCherry/mat15

mat67/mat67; sqh-sqh:mCherry/UASp-shg:GFP

shg:GFP/mat67; sqh-Moe:mCherry/mat15

shg:GFP/+; sqh-Moe:mCherry, mat15/TRiP

mat67/mat67; sqh-Moe:mCherry/UASp-shg:GFP

Fly stocks for rescue experiments. To test RNAi specificity and *shg* transgene function, the following flies were generated and studied. Embryos were generated at 23°C. To rescue embryos with reduced E-cadherin levels, one maternal copy of a ubi-shg:Venus transgene (a fusion between YFP and *shg* ubiquitously expressed under the ubi promoter) was integrated into landing site attP40 on the second chromosome in *Drosophila* whose E-cadherin levels were reduced by RNAi (Appendix B, Fig. B3a and b). Embryos from the following genotypes were collected and imaged.

> Control 1: ubi-shg:Venus/CyO
>
> Control 2: sqh-gap43:mCherry→attP40/ubi-shg:Venus; TRiP/mat15
>
> Lower E-cadherin: sqh-gap43:mCherry→attP40/+; TRiP/mat15

3.2.2 Time-lapse imaging & tissue elongation measurements

Confocal time-lapse imaging and **tissue elongation measurements** were performed as described in Chapter 2, Section 2.2 Materials and methods.

Stereoscope time-lapse imaging. Embryos aged 2-4 hours were mounted in halocarbon oil 27 (Sigma) on transparent apple juice plates made from agar (Apex). Embryos were lined-up and oriented with the head towards the left and the dorsal side upwards (Fig. 3.1.a) using ultra-sharp Dumont Tweezers (Style 55, Electron Microscopy Sciences). A lateral view of the embryo was imaged on a Leica M165FC stereoscope with a Leica DFC70007 camera with proper contrast/brightness to show the shape of the embryonic tissue. Images were acquired at 1-min time intervals for 3 hours and analyzed manually using ImageJ [149]. To assay tissue elongation and embryo hatching for different genotypes, ~100 embryos were lined up and imaged simultaneously (Appendix B.2, Fig. B1).

3.2.3 Automated image segmentation and analysis

Two-color time-lapse movies of *Drosophila* germband extension were split, projected, and despeckled using ImageJ [149]. Processed movies were segmented and computationally analyzed using the MATLAB-based software SEGGA [79], and errors were corrected manually with the interactive user interface. The tissue elongation starting point $t = 0$ was defined as the time-point when the derivative of the tissue elongation curve intersects zero. Cells were tracked and analyzed between $t = -10$ min and $t = 30$ min for each movie.

To analyze the planar polarized localization patterns of proteins (as described in Chapter 2) in the tissue, average intensities of cell edges (interfaces) at different orientations were measured manually in ImageJ [149]. Contracting edges were edges that fully contacted to a vertex, did not reappear for 10 or more frames, and were at least 6-pixel long at some point before contracting. The contraction rate was calculated as the derivative of the edge length vs. time curve over a 15-frame window. To be included in cell rearrangement analysis, cells must be in the region of interest for at least 5 minutes after $t = 0$. The cell rearrangement rate shown is a uniformly weighted average over 1.5 minutes. Node resolution time was measured as the time difference between when an edge contracted to a vertex and when the two cells initially in contact were no longer in contact, provided that they do not reestablish contact for at least 10 time-frames. All cells at the vertex must be in the region of interest for at least 40 time-frames after vertex formation in order to exclude vertices that quickly leave the field of view.

Methods to analyze the cell shape index p and cell shape alignment Q are described in Chapter 2, Section 2.2 Materials and methods and Appendix A.1 Supplementary materials and methods.

3.2.4 Immunohistochemistry

The primary antibody used for staining E-cadherin was rat anti-DE-cadherin DCAD2, 1:25 (Developmental Studies Hybridoma Bank, DSHB). The goat anti-rat secondary antibody conjugated to Alexa-488 was used at 1:500 (Molecular Probes), and Alexa-647-conjugated phalloidin was used at 1:100 (Molecular Probes). Embryos aged 3-6 hours were fixed 1 hour in 3.7% (vol/vol) formaldehyde (Sigma) in 0.1M sodium phosphate pH 7.4 and heptane. Embryos were mounted in Prolong Gold antifade reagent (Molecular Probes) and imaged on a Zeiss LSM880 laser scanning confocal microscope with a 40X/1.2 NA water-immersion objective. Two-color Z stacks were acquired with the standard PMT detector (Zeiss) at 0.5-μm steps. Maximum intensity projections of 2-3 μm were taken in the apical junctional plane (Fig. 3.2a). E-cadherin and F-actin enrichment at junctions relative to in the medial/cytoplasmic domain was quantified using EPySeg, a tool for epithelia image segmentation using deep learning [84], and custom codes.

3.2.5 Fluorescence recovery after photobleaching (FRAP)

The ventrolateral germband of stage 7 embryos expressing GFP-tagged E-cadherin were imaged at 0.5-s time intervals with a Zeiss LSM880 laser scanning confocal microscope and a 63X/1.4 NA oil-immersion objective. A 1.0-μm-sided square region at the cell cortex was photobleached at the third time point ($t = 0$) (Appendix B2, Fig. B6). The fluorescence intensity in the bleached region was tracked for 100 frames and measured at each time point using ImageJ [149]. Intensities were calibrated by subtracting the postbleaching fluorescence intensity.

3.2.6 Measurement of protein intensity

Manual protein intensity measurement. Protein intensities were manually measured using ImageJ [149]. 3-pixel-wide straight lines were drawn along cell edges to measure junctional protein intensities and inside of cells to measure cytoplasmic protein intensities (Appendix B.2, Fig. B4a and b). The protein intensity reported at each cell edge was the line-averaged pixel intensity along the straight line overlapping with the edge and excluding the nodes (i.e. multicellular junctions). The protein intensity reported for each embryo was the mean value of all cell edge intensities within the imaged region of the germband tissue in that embryo. The protein intensity reported for each genotype was the mean value of all embryo intensities of the same genotype. When measuring planar polarity, cell edges within 30 degrees of the anterior-posterior axis are considered as along the AP axis, while cell edges within 30 degrees of the dorsal-ventral axis are considered as along the DV axis (Appendix B.2, Fig. B4c). Intensities were calibrated by subtracting the cytoplasmic fluorescence from the junctional fluorescence (Appendix B.2, Fig. B2).

Automated protein intensity measurement. We also quantified fluorescence intensities of proteins at cellular junctions and in the cell cytoplasm using a semi-automated image analysis. The results were validated with manual measurements in a subset of images (Appendix B.2, Fig. B4d-g). The automated process relied on the segmentation of cell outlines in the E-cadherin channel, which was then used to measure pixel intensities in each of the channels. We used a pre-trained deep neural network (the Linknet model with a vgg16 backbone in the Python package EPySeg [84]) to generate the segmentation masks, and identified the nodes, edges, and cells of the resulting network [195, 196]. When measuring protein intensities in the cell cytoplasm, we defined a sampling window equivalent to repeated erosion to 1/4 of the cytoplasmic area using the cross structuring element. For each edge, we defined a sampling window that covered one binary dilation of the edge padded with 3 pixels on each end to exclude the nodes (i.e. multicellular junctions). We recorded the mean pixel intensities of all cell edges using both the curved and straight-line approximations of the edge, and used the latter to compare with manual measurements. To reduce the uncertainty in segmentations due to image blur, we manually excluded areas of poor focus from each frame's regions of interest (ROI) after z-projection (Appendix B.2, Fig. B4e and f). We found generally good agreement between automated and manual measurements in 46 frames from different experiments (Appendix B.2, Fig. B4). To study planar polarity, we defined and identified edges along the AP or DV axis as those that were oriented parallel to the AP or DV axis within 30-degree margins (Appendix B.2, Fig. B4).

3.2.7 Western blotting

Embryos aged 3-5 hours were dechorionated for 2 min in 50% (vol/vol) bleach, washed in distilled water, and lysed in protein loading buffer (LI-COR). After blocking at 70℃ for 10 min, protein samples were assayed by Pierce BCA Protein Assay Kit. Lysates from ~15 embryos per lane were run on a mini-protean precast gel (Bio-Rad). Proteins were transferred to a 0.45-μm polyvinylidene fluoride membrane (Bio-Rad) by standard protocols. The primary antibodies were: rat anti-DE-cadherin DCAD2, 1:100 (DSHB); mouse anti-β-tubulin E7, 1:100 (DSHB). Secondary antibodies were rat and mouse IRDye 800CW, 1:5,000 (LI-COR), and were detected by fluorescent imaging (ChemiDoc MP Imaging System, Bio-Rad).

3.3 Results

3.3.1 *Drosophila* germband epithelial tissue convergent extension is influenced by E-cadherin levels

To explore the role of cell-cell adhesion in epithelial tissue morphogenesis, we first investigated how the level of E-cadherin (E-cad[7]), a key molecule mediating cell-cell adhesion, influences *Drosophila* germband epithelial tissue mechanical behaviors *in vivo* before and during body axis elongation (Fig. 3.2). To decrease E-cad levels from wild-type levels in the embryo, transgenic RNAi was used to partially inhibit E-cadherin expression [194]. To increase E-cad from wild-type levels, transgenic E-cad was overexpressed using the Gal4/UAS system (Fig. 3.2a) (See Section 3.2 Materials and methods). As a result, we obtained different groups of embryos with systematically modulated E-cad levels from 0.5 to 3-fold of wild-type E-cad levels (Fig. 3.2b).

During early *Drosophila* embryonic development, the germband epithelial tissue rapidly extends along the anterior-posterior (AP) axis (Fig. 3.1a) to roughly double the length of the head to tail body axis in about 30 min (Fig. 2.4b and Fig. 3.2e). In wild-type *Drosophila* embryos (n = 100 embryos), the germband tissue in most embryos (98%) fully elongates two-fold within 30 minutes after the onset of body axis elongation, and most embryos (95%) hatch to larvae successfully (Fig. 3.2 c-e). In contrast, a significant portion of embryos do not have fully elongated germband tissues or hatch to larvae in the group with lower E-cad levels, and this portion is bigger when E-cad levels are further decreased (Fig. 3.2 c and d). The defects in tissue elongation and embryo hatching were successfully rescued by expressing transgenic E-cad in these embryos (Appendix B, Fig. B3c and d), indicating that E-cad was the dominant protein influenced by the RNAi. Interestingly, increasing E-cad levels compared to wild-type produces more embryos

---

[7] For simplicity, E-cadherin is abbreviated as 'E-cad' in the main context of Chapter 3.

without fully-elongated germband tissues or that failed to hatch (Fig. 3.2 c and d). These results

indicate that both decreasing and increasing E-cad levels compared to wild-type causes abnormal

tissue behaviors and developmental defects in embryos. However, decreasing E-cad levels by 50%

produces a much higher percentage of embryos with defects in tissue elongation (72%) and

unhatched embryos (83%) than increasing E-cad levels by 300% (25% and 32%, respectively)

(Fig. 3.2 b-d), suggesting that the mechanisms underlying how E-cad levels modulate epithelial

tissue mechanical behaviors may be complex. Meanwhile, because strictly increasing or

decreasing E-cad levels lowers the chance of full body axis elongation and hatching of embryos

(Fig. 3.2 c and d), it appears there is an optimized range of E-cad levels for tissues to achieve their

target morphogenetic movements.

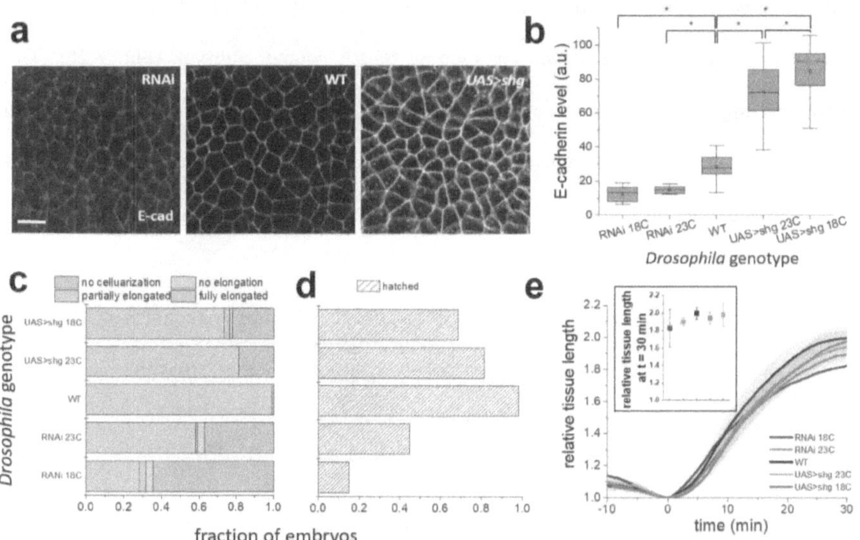

Figure 3.2. E-cadherin levels influence *Drosophila* development, including convergent extension of the germband epithelium during body axis elongation. (a) Confocal images from fixed *Drosophila* germband tissues with different levels of E-cadherin expression. The expression of E-cadherin, labeled by GFP, is decreased from WT (middle) by RNAi (left) and increased from

WT by transgenic overexpression (right). Scale bar, 10 μm. (b) E-cadherin levels in the germband tissue of embryos of different genotypes are quantified in fixed samples. RNAi at 18°C decreases E-cadherin to half the level of wild-type, and one copy of E-cadherin transgene (UAS>shg) at 18°C triples the E-cadherin level relative to wild-type. n = 7-41 fixed embryos. (c) and (d), n = 100 embryos. (c) Body axis elongation and (d) embryo hatching are influenced by E-cadherin levels. The more the E-cad level is increased by transgenic overexpression (UAS>shg) or decreased by RNAi (TRiP), the fewer embryos elongate their body axis fully (D) and hatch (E) compared with embryos of controls. (e) In *Drosophila* embryos that undergo body axis elongation, the epithelial tissue rapidly doubles in length in about 30 min. The E-cad levels in this study do not significantly affect the extent of tissue elongation at t = 30 min for embryos that reach this stage of development (Inset). n = 5-8 embryos. Error bars, standard deviation. For time axes, t = 0 is set as the time when body axis elongation starts unless otherwise stated.

Surprisingly, so long as body axis elongation initiates at $t = 0$, on average, the germband tissues from embryos with different E-cad levels display similar elongation trajectories and final relative lengths at 30 minutes after the onset of body axis elongation (Fig. 3.2e). We do observe that the germband of embryos with either lower or higher E-cad levels are slightly shorter than in wild-type embryos and have larger variation in length among embryos (Fig. 3.2e). These observations might reflect that morphogenetic processes *in vivo* are relatively tolerant of absolute changes in E-cadherin levels due to compensatory strategies. This is potentially consistent with previous findings that differences in cohesion induce cell sorting *in vitro* and *ex vivo* but not *in vivo* [185, 197].

To further investigate the mechanisms of how E-cad influences epithelial tissue mechanics and morphogenesis, we need to take a deeper look at cellular-level behaviors within the germband tissues of embryos with perturbed levels of E-cad.

3.3.2 Cell rearrangements are influenced by E-cadherin levels in *Drosophila* embryos

Tissue remodeling and flow during *Drosophila* body axis elongation are driven mainly by oriented local cell rearrangements [41, 79, 103, 126]. To understand how cell-cell adhesion influences germband tissue morphogenesis, we investigated how E-cad levels affect cell rearrangement. A typical cell rearrangement in the germband tissue during body axis elongation is initiated by contraction of a cell-cell contact along the dorsal-ventral (DV) axis, bringing cells together at a high-order vertex. This vertex then resolves to form a new contact between cells along the anterior-posterior (AP) axis (Fig. 3.3a). The total number of cell rearrangements occurring in the germband tissue increases when tissue elongation starts and ceases when fast tissue elongation ends. Through oriented cell rearrangements, the germband tissue extends along the AP axis while narrowing in the DV direction (Fig. 3.3a). Cell rearrangements are associated with cell junction disassembly and assembly, including junctional E-cad molecule remodeling and transportation. Therefore, less E-cad in the tissue might potentially speed up cell rearrangement, as less E-cad needs to be disassembled and removed during junctional remodeling, while more E-cad might slow down cell rearrangement. To study the effects of E-cad levels on cell rearrangement, we first quantified cell rearrangement speed. We analyzed the average contraction rate of shrinking cell-cell contacts before vertex formation, measured the lifetime of vertices formed during tissue remodeling, and calculated the average growth rate of new horizontal contacts between cells during vertex resolution. We found the dominant effect of varying E-cad levels was on vertex life time.

Interestingly, we found that cell rearrangement speed quantified by vertex life time exhibits a biphasic dependence on E-cad levels (Fig. 3.3b). Reducing E-cad from normal levels produces faster rearrangement. In contrast, small increases in E-cad above normal levels produces slower rearrangement. Surprisingly, strongly increased E-cad levels result in faster rearrangement compared to controls with normal levels of E-cad. This biphasic dependence on E-cad levels is not consistent with predictions of the simple model that increased adhesion would result in slower rearrangement associated with additional time to disassemble cell junctions as cells exchange neighbors. Instead, very high or low E-cad levels in the germband tissue produce faster cell rearrangements, indicating there might be lower barriers to cell rearrangement and that the tissue might be more fluid-like.

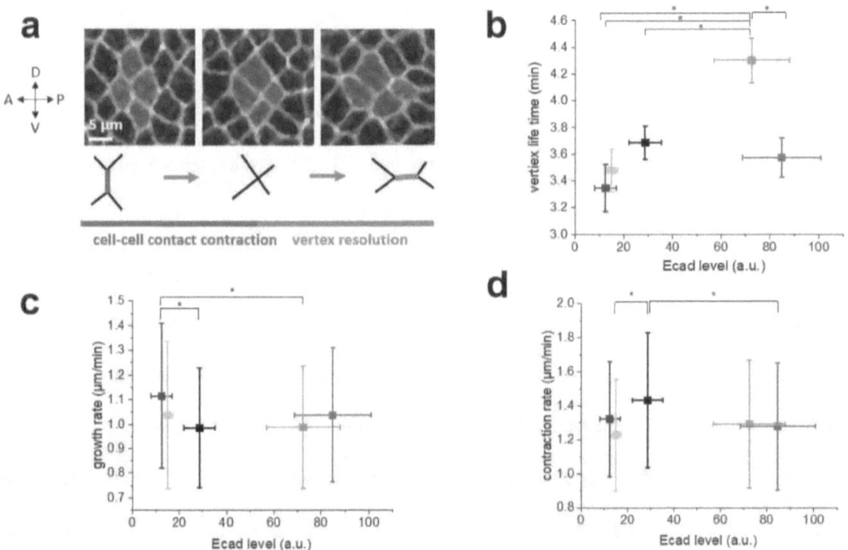

Figure 3.3. Cell rearrangements are influenced by E-cadherin levels in *Drosophila* embryos. (a) Cell rearrangements in the *Drosophila* embryonic epithelium. Top: Images from a confocal time-lapse movie tracking a cell rearrangement process. Bottom: Cell rearrangement schematic. In the first step, rearrangement is initiated by contraction of cell-cell-contacts, bringing cells together at a high-order vertex. In the second step, vertices are resolved by the formation of new contacts between cells. The time period that cells stay at the vertex is called vertex life time. (b) Cell rearrangement speed quantified by vertex life time is modulated by E-cadherin levels in a biphasic manner. Both high and low E-cad levels produce shorter vertex life time indicating faster cell rearrangement and more fluid-like tissue behaviors. (c) The rate of new cell edge growth during rearrangement. Lower E-cadherin levels appear to slightly speed up new cell edge growth. (d) The contraction rate of contracting cell edges during rearrangement. Increasing or decreasing E-cadherin levels appear to slightly slow down cell edge contraction. Unless otherwise specified, in (b)-(d) and other figures of Chapter 3: Blue, RNAi 18C; Cyan, RNAi 23C; Black, WT; Orange, UAS>shg 23C; Red, UAS>shg 18C.

A recent study showed that cell edge growth along the AP axis helps drive vertex resolution and is correlated with the magnitudes of imbalanced forces generated within cells at the ends of the new edge along the AP axis [152]. When we measured the rate of cell-cell contact growth along the AP axis, we found that increased E-cad levels have no effect on the growth rate while decreased E-cad levels appear to slightly accelerate growth (Fig. 3.3c). This suggests that the shorter vertex life time in embryos with lower E-cad levels might be related to their faster cell edge growth and potentially stronger imbalanced forces along the AP axis. Interestingly, the rate of cell-cell contact contraction along the DV axis displays an opposite result – both decreasing and increasing E-cad levels can slightly slow down this step of cell rearrangement (Fig. 3.3d).

The overall tissue elongation rate and extent among different E-cad level groups are similar (Fig. 3.2e), suggesting there may be complementary mechanisms to balance the modest differences we observe in cell rearrangement speeds.

3.3.3 Cell patterns are modulated by E-cadherin levels in *Drosophila* embryos

The biphasic dependence of cell rearrangement speed on E-cad levels in the germband tissue might reflect changes of mechanical force balances on cells discussed above, or a change in tissue mechanical properties that control how cells move in response to forces. Next, we investigated if and how tissue mechanical properties are modulated by E-cad levels. The work in Chapter 2 combining experiments and simulations suggests that solid-like vs. fluid-like behavior of the germband epithelial tissue can be predicted by two metrics of cell patterns – the cell shape index and the cell alignment index, which can be calculated from cell shapes in the tissue. We speculated that changes in cell-cell adhesion might result in cell shape changes that could influence tissue mechanics. As cells tend to increase cell-cell contact lengths in the presence of adhesion molecules, we predicted that increased E-cad levels would result in longer cell contacts and thus longer cell perimeters. To quantify cell shapes in the *Drosophila* germband tissue, we acquired time-lapse movies of embryos with cell membranes tagged by fluorescent proteins. We segmented and analyzed the resulting movies to quantify the cell shape index p, which is the ratio of cell perimeter P to the square root of cell area A ($p = P/\sqrt{A}$, also see Chapter 2 and Appendix A), cell shape alignment Q, which represents tissue anisotropy measured by the triangular method (Fig. 2.3 and Fig. 3.4a) (details in Chapter 2 and Appendix A), and vertex coordination number z, which is a metric for cell packing disorder.

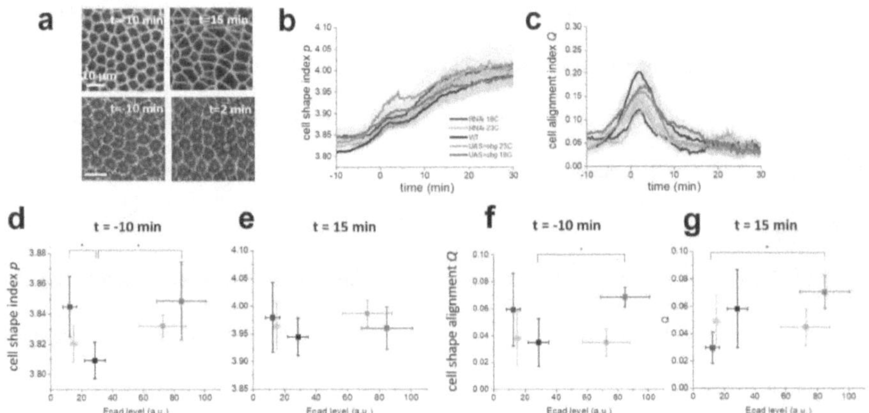

Figure 3.4. Cell patterns are modulated by E-cadherin levels in *Drosophila* germband tissues.
(a) Confocal images of epithelial tissue in developing *Drosophila* embryos. Top: Segmentation of epithelial cells at 10 minutes before and 15 minutes after the onset of body axis elongation with the SEGGA software package [79]. Overlaid polygon representations (green lines) are used to quantify cell shapes. Bottom: Quantification of cell shapes at 10 minutes before and 2 minutes after the onset of body axis elongation by the triangle method (Section 2.2.1.5 and Appendix A.1.1). Cell centers (green dots) are connected with each other by a triangular network (red bonds) to quantify tissue anisotropy. Scale bar, 10 μm. (b) Cell shape index p increases with time during axis elongation in embryos with different E-cadherin levels. There is a small peak at around $t = 2$ min when the ventral furrow is pulling the tissue. Wild-type embryos seem to have relatively lower p than other groups. (c) Cell shape alignment Q first increases and then decreases with time during axis elongation, which peaks at around $t = 2$ min in embryos with different E-cadherin

levels. Wild-type embryos seem to have relatively higher Q around the peak. (d) Cell shape index p shows a biphasic dependence on E-cadherin levels at 10 minutes before the axis elongation, but (e) this dependence disappears at $t = 15$ min when massive cell rearrangements happen. (f) and (g) Cell shape alignment Q shows less dependence on E-cadherin levels. (f) At 10 minutes before the axis elongation, Q shows similar relationship with E-cadherin levels as p partially because Q contains information of cell shape change. Q combines information about both cell shape Q_s and cell alignment Q_a. Q_a which excludes the influence of p, is actually independent with E-cad levels (Appendix B2, Fig. B5). (g) Cell shape alignment Q and Q_a (Appendix B2, Fig. B5) increases with E-cadherin levels at $t = 15$ min while p and Q_s (Appendix B2, Fig. B5) is independent with E-cad levels.

In wild-type *Drosophila* embryos, cells take on nearly isotropic hexagonal shapes at 10 minutes before body axis elongation when the tissue is static and solid-like, but change shape dramatically during body axis elongation. Consistent with this, the cell shape index p increases during axis elongation in these embryos. We found that cell patterns in embryos with different E-cad levels show a similar trend to those in wild-type embryos. In germband tissues with different E-cad levels, the cell shape index p increases dramatically with time (Fig. 3.4b) while the cell alignment Q first increases and then decreases showing a peak at around t = 2 min (Fig. 3.4c), consistent with observations in Chapter 2. Cells take on nearly isotropic hexagonal shapes before tissue remodeling ($t = $ -10 min) and become more elongated with higher p during tissue remodeling ($t > 0$) when internal and external mechanical forces anisotropically deform cells (Fig. 3.4a). The elongations of cells are aligned with the DV axis at the onset of tissue remodeling (Fig. 3.4a) and become randomly oriented when tissue elongation continues while Q decreases. We then examined how changes in E-cad levels influenced cell shapes (p and Q) prior to the onset of and during cell rearrangement and tissue remodeling.

We first examined the influence of E-cad levels on the tissue when it is static and solid-like, before remodeling begins. In embryos with normal E-cad levels, cell shapes at $t = $ -10 min are close to that predicted at the solid-fluid transition point in the isotropic vertex model (Fig. 3.4d) [67] and Q is close to zero, suggesting the tissue is isotropic. Remarkably, we found a biphasic dependence of the cell shape index p on E-cad levels. Tissues with either increased or decreased E-cad levels display increased p compared to tissues in normal control embryos (Fig. 3.4d). This behavior is not consistent with our prediction that increased adhesion would result in longer cell contacts and increased p. Interestingly, however, we note that tissues comprising cells with the longest perimeters (Fig. 3.4d, high p) tend to rearrange fastest (Fig. 3.3b, short lifetimes), suggesting a potential link between cell shapes and cell rearrangements. In contrast, when the tissue becomes dynamic and starts remodeling at $t = 2$ min (data not shown) and continues remodeling at 15 min, p no longer varies among embryos with different E-cad levels (Fig. 3.4e).

The cell shape alignment Q of tissues in embryos with different E-cad levels shows a similar dependence on E-cad levels as cell shape index p does (Fig. 3.4f), which is explained by the contribution of cell shape index p (Appendix B2, Fig. B5). However, Q also shows a difference at later times ($t = 15$ min) close to the peak of the cell rearrangement rate, while p is independent of E-cad levels. The cell shape alignment Q increases when E-cad levels increase, but the values are much lower than the peak values of Q at around $t = 2$ min (Fig. 3.4g). As discussed in Appendix A.1.1, the cell shape alignment $Q = Q_s Q_a$ combines information about both cell shape anisotropy Q_s and cell shape alignment Q_a. When we further analyzed Q_s and Q_a (See Appendix A.1.1, Appendix B2, Fig. B5), Q_s is related to the cell shape index p, so Q_s shows the same biphasic dependence on E-cad levels as p, while Q_a is independent of E-cad levels (Appendix B2, Fig. B5a and d), indicating that the difference of Q at t = -10 min is caused by the differences in cell shape anisotropy. At $t = 15$ min, Q_s becomes independent of E-cad levels, similar to p, while Q_a dominates the relationship of Q and E-cad (Appendix B2, Fig. B5b and e), suggesting that the difference of Q at t = 15 min is caused by cell shape alignment. Interestingly, at $t = 2$ min when Q

reaches the peak, E-cad levels do not influence Q values significantly (Appendix B2, Fig. B5c and f).

3.3.4 E-cadherin levels tune the predicted fluidity of the germband epithelium

In Section 3.3.2, we showed that germband tissues with very high or low E-cad levels have faster cell rearrangements, which is typically associated with more fluid-like tissue behaviors. To test if E-cad levels tune tissue mechanical properties, we leveraged the methods developed in Chapter 2 using the cell shape index p, cell shape alignment Q, and vertex coordination number z as metrics for cell patterns (detailed description and calculation in Chapter 2 and Appendix A) to analyze tissue fluidity of the germband. If we set the y-axis as $\overline{p}_{corr}$, the average cell shape index corrected by z as $\overline{p}_{corr} = \overline{p} - (z - 3)/B$, where $B = 3.85$, and the x axis as the cell shape alignment Q (Fig. 3.5), according to Chapter 2, the predicted solid-fluid transition line (Fig. 3.5, solid lines) is $\overline{p}_{corr} = 3.818 + 4bQ^2$, where $b = 0.43$. If the tissue sits above the line, it is predicted to behave as a fluid-like material, while the tissue is predicted to be solid-like if it is below the line.

We first look at the tissue 10 minutes before body axis elongation when the tissue is static. For wild-type embryos, the tissue is in the solid-like region at this time (Fig. 3.5a black). When E-cad levels are decreased, the cell patterns change and the tissue moves towards the transition line and is predicted to become more fluid-like (Fig. 3.5a blue and cyan). This relationship between tissue fluidity and E-cad levels is consistent with a simple, intuitive hypothesis that tissues with less E-cad are less adhesive and more fluid-like. Interestingly, when E-cad levels are increased, again the tissue moves towards the transition line and is predicted to become more fluid-like, which is contrary to the hypothesis (Fig. 3.5, orange and red). Therefore, both higher or lower E-cad levels produce tissues predicted to be more fluid-like – that is, the tissue fluidity also has a biphasic

dependence on E-cad levels, which is consistent with the biphasic relationship between cell

rearrangement speeds and E-cad levels. We observed a similar relationship 15 minutes after body

axis elongation, when the tissue is dynamic (Fig. 3.5 b), although the tissue trajectories through

the p-Q parameter space are somewhat different than those in static tissues.

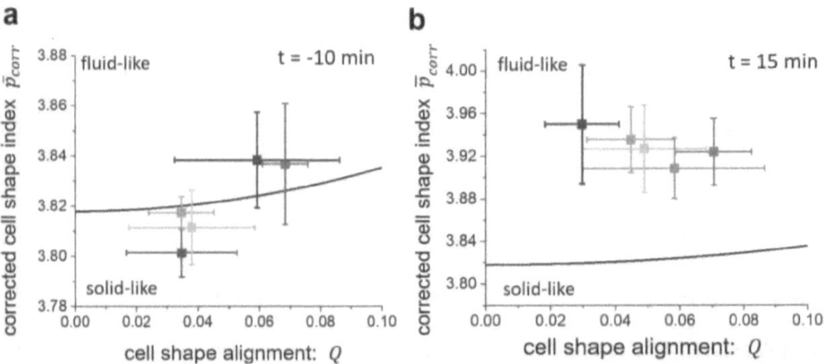

Figure 3.5. Epithelial tissue fluidity is tuned by E-cadherin levels in *Drosophila* embryos. (a) and (b), Combining $\bar{p}_{corr}$, which is the average cell shape index corrected by vertex coordination number z, and cell shape alignment Q can be used to predict epithelial tissue fluidity. The solid-line, $\bar{p}_{corr} = 3.818 + 4bQ^2$, is the predicted tissue solid-fluid transition line in the anisotropic vertex model in Chapter 2. n = 5-8 embryos. (a) The fluidity of the germband epithelial tissue has a biphasic dependence on E-cadherin levels at 10 minutes before body axis elongation. For wild-type embryos, the tissue stays in the solid-like zone (black). Both decreased (cyan) and increased (orange) E-cadherin levels result in the tissue moving closer to the transition line. Further decreasing (blue) and increasing (red) E-cadherin result in a tissue predicted to be even more fluid-like, even crossing the solid-fluid transition line. (b) At 15 minutes after the onset of body axis elongation, when there are cell rearrangements, the fluidity of the germband epithelial tissue also appears to become more fluid-like when E-cadherin is either increased (orange, red) or decreased (cyan, blue) compared to wild-type (black), as tissues move further into the predicted fluid-like region of the parameter space

Both the cell rearrangement speed and the tissue fluidity prediction from cell patterns

suggest that higher or lower E-cad levels produce more fluid-like tissues, contradictory to most

simple models of cell-cell adhesion. Local remodeling of junctions at cell-cell contacts during epithelial morphogenesis has recently been shown to involve complex mechanosensitive regulatory mechanisms [95, 159, 198]. For example, E-cad molecules are coupled with the actomyosin cytoskeleton to organize and transmit tensions and their turnover and trafficking are thought to be related to junctional remodeling [95, 159, 198]. To explore potential mechanisms underlying the biphasic dependence of tissue fluidity on E-cad levels, next, we explore how E-cad levels influence cytoskeletal protein localization and dynamics.

3.3.5 E-cadherin protein turnover dynamics, junctional myosin II levels and cell area fluctuations are modulated by E-cadherin levels

To further explore the mechanisms underlying the dependence of tissue fluidity on E-cad levels, we first investigated E-cad protein turnover dynamics using fluorescence recovery after photobleaching (FRAP) (See Section 3.2 Materials and methods and Appendix B.2, Fig. B6). Here we use the characteristic time and the mobile fraction of E-cad to quantify junctional E-cad dynamics (Fig. 3.6a). Considering the planar polarized localization pattern of key proteins in the germband tissue, we conducted FRAP experiments on cell edges (bonds) that are oriented either along the AP or DV axis and measured the characteristic time of junctional E-cad recovery and the mobile fraction of E-cad molecules during FRAP. Differences in FRAP results from cell bonds along the AP or DV axes were not observed (data not shown), so we analyzed junctional E-cad dynamics regardless of bond orientation. We found that characteristic times of E-cad recovery are not significantly different in embryos with different E-cad levels (Fig. 3.6b). Since junctional E-cad transportation is dominated by endocytosis during cell rearrangement, this indicates that E-cad levels may not modulate endocytosis directly. However, when E-cad levels are decreased compared to wild-type, the mobile fraction of E-cad at cell-cell contacts decreases (Fig. 3.6c). A recent study found that the mobile E-cad protects short cell-cell contacts from collapsing because

these borders may require more mobile E-cad to maintain them [192]. When levels of mobile E-

cad were reduced in the *Drosophila* embryo at a later stage of development, formation of

multicellular rosettes in epidermal cells is increased and cell motility is increased [192]. Therefore,

the decreased mobile fraction of E-cad in tissues with lower E-cad levels suggest that cell

rearrangements happen more easily and cell motility is increased. Meanwhile, less E-cad makes

the tissue less adhesive and easier to remodel. These might explain why lower E-cad levels produce

faster cell rearrangement and a more fluid-like tissue.

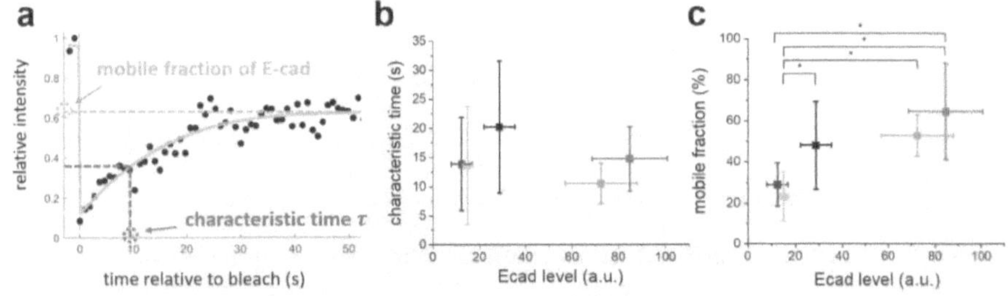

Figure 3.6. Junctional E-cadherin protein dynamics are influenced by E-cadherin levels. (a) A typical plot of relative fluorescence intensity after photobleaching of E-cad:GFP in the germband tissue during *Drosophila* body axis elongation. Each black dot represents a single time point. Bleaching occurs at t = 0. Then the relative intensity drops from ~1 to almost 0 before recovering. The maximum relative intensity recovered is from the mobile E-cadherin population and called the mobile fraction. The recovery curve is fitted by an exponential function, and the exponential constant is the characteristic time τ. (b) Characteristic time of junctional E-cadherin turnover is independent of E-cadherin levels. (c) Mobile fraction of junctional E-cadherin is lower in tissues with lower E-cadherin levels, indicating that there are less mobile E-cadherin proteins when total E-cadherin is decreased. (b) and (c), n = 6 embryos.

E-cadherin-based cell-cell adhesion and the actomyosin networks are tightly coupled.

Myosin II and actomyosin contractility have been reported to modulate E-cad behavior [183-192].

Recent work shows that E-cad plays an active role in regulating myosin II and tensions [159, 193].

We next explored if E-cad levels tune tissue fluidity by actively modulating myosin II and contractile tensions in this tissue. There are two pools of apical myosin in the germband tissue [160]. Junctional myosin II is enriched at vertical cell-cell contacts at the level of adherens junctions during germband extension and generates anisotropic forces that drive cell-cell contacts oriented along the DV axis to shrink [160]. In contrast, medial myosin II generates pulsed contractions driving apical cell area fluctuations, which might provide active fluctuations to promote cells to overcome energy barriers and change their neighbors [160].

We measured both junctional and medial myosin II and F-actin levels in embryos with different E-cad levels. We found that when E-cad levels are increased (Fig. 3.7a), F-actin levels show no change, but junctional myosin II levels are decreased (Fig. 3.7b). This is consistent with previous reports [193]. Decreased myosin would likely lead to decreased cortical tensions at cell-cell contacts around the perimeter of cells. In the vertex model, this would be associated with lower energy barriers to cell rearrangement and more fluid-like tissue behavior [67]. This coupling between the adhesion machinery and the contractile tension machinery may help explain why we observe that tissues with higher E-cad levels appear more fluid-like.

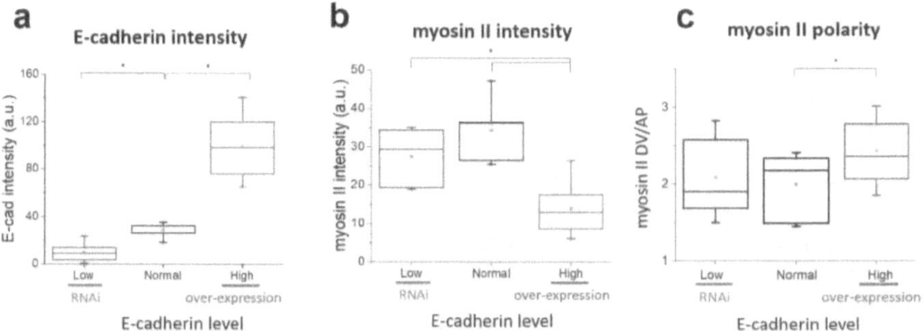

Figure 3.7. Junctional myosin II is modulated by E-cadherin levels. Measurements of intensities of (a) E-cadherin marked by GFP and (b)-(c) Myosin II tagged by mCherry from live confocal movies. n = 11-18 embryos. (a) E-cadherin level is decreased to 1/3 of the wild-type level by RNAi and increased to 3 times of the wild-type level by E-cad transgene overexpression (See 3.2

Materials and methods). (b) Before the onset of body axis elongation, myosin II levels are decreased in embryos with higher E-cadherin levels. (c) During body axis elongation, the polarity of myosin II, quantified by the mean myosin II intensity of cell edges oriented along the DV axis over that along the AP axis, is increased in embryos with higher E-cadherin levels.

Since the germband tissue is highly anisotropic, we also analyzed if E-cad levels influence myosin II planar polarity, which is thought to drive oriented cell rearrangements [40, 103, 104, 126-129]. We found that when E-cad levels are increased (Fig. 3.7a), the myosin II planar polarity is slightly increased (Fig. 3.7a). More specifically, there is more myosin II enriched at cell junctions oriented along the DV axis compared to those oriented along the AP axis, potentially generating relatively higher internal driving forces for cell rearrangement and tissue remodeling. This could also help explain why tissues with higher E-cad levels display faster cell rearrangement.

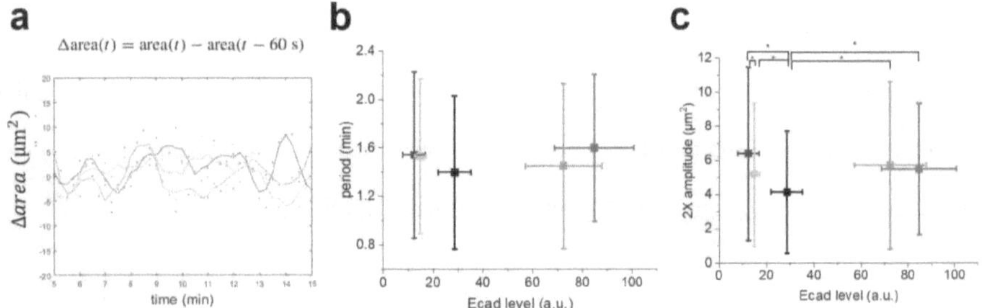

Figure 3.8. Apical cell area fluctuations are affected by E-cadherin levels. Individual cells in the germband with different E-cadherin levels are tracked and their apical areas are measured. (a) Examples of fitted cell area fluctuations over time. Area change is the difference in cell area compared to the value 60 seconds ago [199]. (b) Cell area fluctuation period is not influenced by E-cadherin levels. n = 283-367 area fluctuation cycles. (c) The amplitude of cell area fluctuation shows a biphasic dependence on E-cadherin levels. When E-cadherin levels are higher or lower than wild-type, cells fluctuate more. n = 482-516 area fluctuation cycles.

Next, we measured medial myosin II levels and apical cell area fluctuations that are thought to result from medial myosin contractility (Fig. 3.8a) [199]. We found no difference in medial myosin levels in embryos with different E-cad levels. We did not observe changes in the period of apical cell area fluctuations (Fig. 3.8b). However, the amplitudes of cell area fluctuations increase in embryos with either higher or lower E-cad levels than wild-type (Fig. 3.8c). Active cell area fluctuations have been proposed to contribute to cells overcoming energy barriers to rearrangements and behaving in a more fluid-like manner [69]. The cell area fluctuation amplitude is partially determined by the imbalanced forces among cells within the tissue [152]. Since actomyosin networks in cells are coupled to each other through E-cad, abnormal E-cad levels may cause larger imbalances of forces, making cells fluctuate more dramatically. These results reveal a role for E-cad in regulating myosin II localization and contractile activity in cells, providing a potential mechanistic linkage between tissue fluidity and E-cad levels.

3.4 Discussion

In this work, we experimentally investigate the role of cell-cell adhesion in epithelial mechanics and morphogenesis *in vivo*. We find that modulating overall E-cad levels influences cell rearrangement, cell patterns, and tissue fluidity in the germband epithelium during *Drosophila* body axis elongation. We find a biphasic dependence of cell rearrangement speed on E-cad levels, where cells rearrange faster in embryos with either lower or higher E-cad levels than wild-type embryos, suggesting more fluid-like tissue behaviors in embryos with perturbed E-cad levels. Meanwhile, when E-cad levels are increased or decreased from the wild-type level, two cell pattern metrics, the cell shape index p and cell shape alignment Q, change (Figure 3.4) in a manner such that the tissue is predicted to behave more fluid-like based on the anisotropic vertex model analyses from Chapter 2 (Figure 3.5). In more fluid-like tissues, the energy barrier for exchanging cell neighbors is low, which could contribute to faster cell rearrangements. These biphasic results reveal surprising links among cell rearrangement, cell shape, tissue fluidity, and cell-cell adhesion. In particular, we find that tissues with both higher and lower E-cad levels behave more fluid-like with lower energy barriers to cell rearrangement and faster cell rearrangements. We further investigate molecular and mechanical mechanisms underlying this linkage by analyzing E-cad turnover dynamics, actomyosin localization patterns, and apical cell area fluctuations in embryos with different E-cad levels. We find that the mobile fraction of E-cad decreases in embryos with lower E-cad levels; the overall junctional myosin II level decreases, but the polarity of myosin II slightly increases in embryos with increased E-cad levels; and the cell area fluctuation shows increased magnitude in tissues with either higher or lower E-cad levels.

These findings on molecular and cellular behaviors provide insights into potential mechanisms underlying how E-cadherin enacts its dual roles to tune the mechanical behaviors of epithelial tissues *in vivo*. First, when the overall E-cad level is decreased, the FRAP results show that the fraction of mobile E-cad decreases, indicating potentially higher cell motility in Ref. [192] because there is less mobile E-cad to protect short cell-cell contacts from collapsing. This higher cell motility might contribute to more fluid-like tissue behavior and easier cell rearrangements. Therefore, tissues with lower E-cad levels become more fluid-like. Second, when the overall E-cad level is increased, the junctional myosin II level is decreased, which is associated with lower energy barriers to cell rearrangement and more fluid-like tissue behavior in the vertex model predictions [67]. Meanwhile, myosin II planar polarity is slightly increased with relatively more myosin II enriched at cell edges oriented along the DV axis compared to those oriented along the AP axis, potentially generating relatively higher internal driving forces for cell rearrangement and producing more fluid-like tissue behaviors. Therefore, tissues with higher E-cad levels are also more fluid-like.

Remarkably, apical cell area fluctuations also show a biphasic response to E-cad levels: apical cell areas fluctuate more strongly with higher amplitudes in embryos with either lower or higher E-cad levels compared to wild-type. A recent study shows a correlation between periodic contractions of cells and the new cell edge growth, which helps resolve the higher-order vertex formed during cell rearrangement, and proposes that the vertex resolution and new edge growth are related to the imbalance of forces generated in the surrounding cells [152]. Consistent with these results, in embryos with lower or higher E-cad levels, tissues comprising cells with higher-amplitude periodic contractions show shorter vertex life times and slightly higher new cell edge growth rates during cell rearrangements, both of which contribute to more fluid-like tissue behaviors. Moreover, forces generated by cell area fluctuations are thought to be one of the energetic contributions for cells to exchange their neighbors [67, 69]. The stronger apical cell area fluctuations in embryos with either lower or higher E-cad levels might provide more energy for cells to overcome energy barriers and exchange their neighbors, which would contribute to more fluid-like tissue behaviors.

Interestingly, although our findings show that this epithelial tissue becomes more fluid-like when the E-cad level is either increased or decreased, the underlying mechanisms are distinct between the lower and higher E-cad situations, indicating the complex role of cell-cell adhesion in controlling epithelial tissue mechanics and morphogenesis. Moreover, compared with the influence of force-generating proteins such as myosin II, the modulation of tissue behaviors by cell-cell adhesion seems to be subtler and finer. For example, the effects of E-cad levels on tissue mechanical behaviors are more obvious and noteworthy when the tissue is static, such as the germband tissue at 10 minutes before the onset of tissue elongation, but at 15 minutes after the elongation starts, when actomyosin contractility dominates the tissue behaviors, effects from E-cad levels become less significant. Therefore, E-cadherin-based cell-cell adhesion might act as a subtle regulator for living tissues to control their morphogenesis.

In addition, there are some notable observations relating to how cell-cell adhesion controls epithelial tissue morphogenesis. First, *in vivo* morphogenetic processes in living organisms seem to have a relatively high tolerance of overall changes in E-cad. This might be because in epithelial tissues cell-cell adhesion provides passive adhesive forces and domains for actomyosin networks to bind. As a result, cells may only need a threshold amount of adhesion proteins for force-generating proteins such as myosin II to function. However, there appears to be an optimal range of amounts of E-cad for the tissue to function, since both increasing and decreasing E-cad levels from wild-type levels often result in the same tissue mechanical behaviors. Second, *in vivo* tissues seem to have compensatory mechanisms to balance potential defects to promote robustness. For example, even though the molecular, cellular, and tissue-level behaviors are all tuned by E-cad levels, overall tissue elongation remains relatively stable among embryos with different E-cad levels. It will be interesting to study the compensatory mechanisms in the *Drosophila* germband epithelium in the future.

Moving forward, more work needs to be done to further explore the mechanisms by which E-cad regulates actomyosin. Instead of modulating E-cad levels globally, it will also be interesting to locally modulate E-cad molecules and study the effects on tissue mechanical behaviors. Incorporating our results on the effects of cell-cell adhesion on epithelial cell behaviors into more sophisticated models could provide more precise predictions of tissue morphogenetic processes. For example, current models of epithelial tissues such as vertex models often lack the temporal dimension, and our work will be useful for adding in timescales for these models.

Chapter 4. Extending Analyses of Tissue Mechanics in New Directions --- Tissue Mechanics in Three Dimensions and the Inner Ear Round Window Membrane

4.1 Introduction

In Chapters 2 & 3, we developed a framework to explore mechanics in epithelial tissues *in vivo* and investigated the mechanical and molecular mechanisms of epithelial mechanics during convergent extension. I developed and used experimental approaches to probe, manipulate, and analyze mechanical behaviors in the *Drosophila* germband epithelial tissue; I worked with my theory and modeling collaborators to develop an easy, practical and appealing approach to predict epithelial tissue mechanical behaviors *in vivo*.

We now have powerful approaches to study mechanics in living tissues. To show the potential of this research framework, in Chapter 4, I expand my work on epithelial tissues from 2-dimensional to 3-dimensional[8] analyses and apply my experimental approaches to other tissue types.

Section 4.2 presents my current research progress on investigating mechanical behaviors in 3-D of the *Drosophila* germband epithelium by combining high-depth live confocal imaging, quantitative image analysis, and vertex model predictions. Preliminary results are shown and analyzed. In Section 4.3, I contributed my confocal imaging expertise developed in Chapters 2 and 3 to our collaborators' efforts to probe and analyze mechanics in the round window membrane

---

[8] For simplicity, 2, 3-dimensional are abbreviated as 2, 3-D in this book.

(RWM)[9] in the guinea pig inner ear and develop a new drug delivery method into the inner ear by perforation via ultra-sharp microneedles. Results I contributed to this project are presented.

The successful applications of the research methods and ideas developed in the apical junctional *Drosophila* germband epithelial tissue to studying 3-D tissue mechanics and RWM mechanics extend the horizons of my research and show the wider impact on the field of biomechanics of the knowledge and approaches developed in this book.

4.2 Exploring epithelial tissue mechanics in 3-D *in vivo*

4.2.1 Introduction

During morphogenesis, tissues undergo complex morphogenetic processes resulting from mechanics that evolve over time and in 3-D space. Biological tissues are naturally in 3-D, involving spatially distributed molecules, locally regulated subcellular behaviors, and highly curved tissue environments (Fig. 4.1). Indeed, in the early *Drosophila* embryo, for example, just before body axis elongation (Fig. 2.1a and Fig. 3.1a), the germband epithelial cells not only have a diameter of about 10 μm (Fig. 4.1a and b, left side), but also reach a depth of about 30 μm along the apical-basal axis (Fig. 4.1a and b, right side). However, research on the germband epithelial tissues has mainly focused on a region that is a few micrometers deep from the most apical side, in large part because myosin II and E-cadherin molecules are mainly distributed in a narrow region in this area forming the adherens junctions (Fig. 4.1d) [96]. In reality, there are also complex cytoskeletal systems existing on the basolateral cortex of epithelial cells (Fig. 4.1d) [96], which potentially provide different molecular mechanisms for cellular mechanical behaviors from those in the apical side. Nevertheless, in most experimental and computational studies, epithelial tissues

---

[9] For simplicity, round window membrane is abbreviated as RWM in this book.

are simplified from 3-D to 2-D structures by only considering the apical side [7, 40-42 67-72], partially due to the lack of both live imaging methods to take high-depth images and non-invasive methods to probe mechanics *in vivo*.

Recent studies have highlighted the importance of studying tissue mechanics in 3-D [106-109]. In mouse embryos, for example, a 3-D tissue stiffness gradient has been observed along which cells move to shape the early limb bud [108], and it has been found that cells undergo biased intercalation and exhibit both apical rearrangement and polarized basolateral protrusive activity during the convergent extension of the mouse neural plate [106]. Surprisingly, basolateral protrusion and apical contraction are also observed in the *Drosophila* germband epithelium and thought to cooperatively drive *Drosophila* body axis elongation along with apical junctional behaviors [107] (Fig. 4.1c). Similarly, 3-D geometries have recently been included in computational modeling work, e.g. coupling tissue curvature and thickness and using geometric constraints to alter cell rearrangements [68, 109]. However, even with this progress, few studies focusing on *in vivo* mechanics of epithelial tissues in 3-D have been reported.

In this section, combining high-depth live imaging and quantitative image analysis, I studied mechanical behaviors of the *Drosophila* germband epithelial tissue in 3-D. In particular, I explored how cell packings and predicted tissue mechanical behaviors change along the apical-basal axis in the germband tissue. In this ongoing work, our preliminary results show that cell patterns change in a gradient along the apical to the basolateral side, indicating potentially different mechanisms of controlling cellular behaviors, such as cell rearrangement, at the apical, lateral, and basal sides of cells. Furthermore, using the analysis methods developed in Chapter 2, we predict that there is a variation in tissue fluidity along the apical-basal axis of the tissue. Particularly, the basolateral side of the tissue is predicted to show more solid-like behaviors compared with the apical side, which would likely impact both tissue remodeling and flow in the plane of the tissue as well as out-of-plane motion or bending of the tissue. This ongoing project has the potential to expand my expertise to 3-D tissue mechanics, reveal mechanisms underlying the complex mechanical behaviors of living tissues undergoing morphogenetic events, and provide a foundation to understand how improper regulation of tissue mechanics might be associated with birth defects, aberrant wound healing, and cancer metastasis.

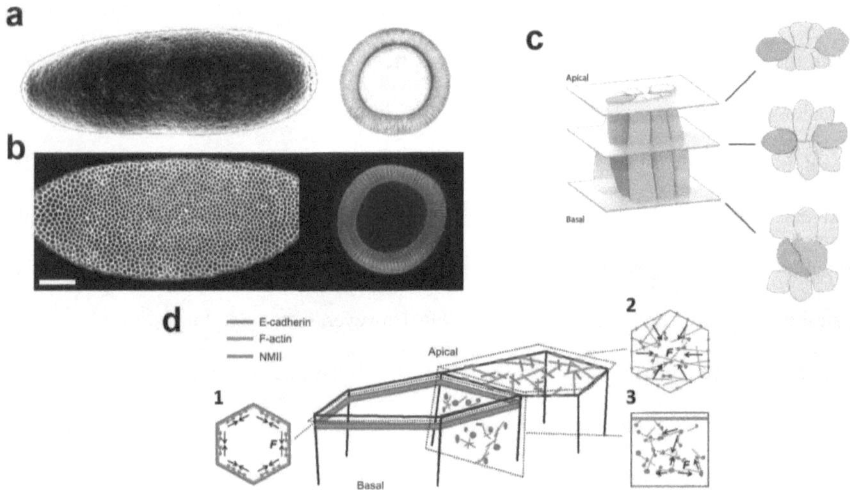

Figure 4.1. 3-D structures and mechanical behaviors in the *Drosophila* germband epithelial tissue. (a) Images showing ventral views of the *Drosophila* embryo surface (left) (Conte et al., 2008, image reprinted by permission from Journal of the Mechanical Behavior of Biomedical Materials) [170, 178] and a 2-D cross-section view of the embryo (right) (Muñoz *et al.*, 2007) [177]. (b) Left, confocal image of the lateral surface of the *Drosophila* embryo with cell membrane marked by mCherry. Image from Marisol Herrera-Perez. Right, confocal image of the cross-section view of the *Drosophila* embryo. Green, cell membrane. Nuclear red, *twist*. Cell surface red, myosin. Image from Eric Wieschaus. (a) and (b), Scale bar, 50 µm. (c) Schematic of cell rearrangement in 3-D during *Drosophila* germband extension. Basolateral protrusion and apical contraction cooperatively drive cell rearrangement in 3-D. Images from mechanobio.info and Ref. [107]. (d) Schematic of cortical cytoskeleton system in the epithelium in 3-D. 1. Circumferential actomyosin cytoskeleton decorating E-cadherin at the *zonula adherens*. 2. Apical actomyosin meshwork. 3.

Cortical F-actin and non-muscle myosin II (NMII) interaction with E-cadherin through the basolateral membrane. Mechanical conditions are different along the apical-basal axis. Mechanical forces (F) are generated by coupling of E-cadherin to the actomyosin cytoskeleton in these different cortical contexts. Image from (Liang et al., 2015) [96].

4.2.2 Materials and methods

In this study, the germband epithelium was investigated in 3-D using *Drosophila* culture and genetics, confocal microscopy, and quantitative image analysis.

Embryos were generated at 23°C and imaged *in vivo* on a Zeiss LSM880 laser-scanning confocal microscope at room temperature. Cell outlines were visualized with gap43:mCherry [144] cell-membrane markers and myosin II molecules were visualized with mCherry fluorescent markers. Time-lapse movies were analyzed with SEGGA software in MATLAB [79] for topologies and dynamics, PIVlab (Version 1.41) in MATLAB [145] for quantifying tissue elongation, and custom code for quantifying cell alignment using the triangle method [130, 146, 147].

Methods for tissue elongation measurement, automated image segmentation, and the cell shape index p and cell shape alignment Q analysis are described in Chapter 2, Section 2.2 Materials and methods and Appendix A.1 Supplementary materials and methods.

Details of other materials and methods are listed below. Unless otherwise noted, error bars are the standard deviation (SD).

4.2.2.1 Fly stocks and genetics

The epithelial tissue studied was the germband tissue in *Drosophila* embryos. Embryos were generated at 23°C and analyzed at room temperature. Embryos were *yw* with double maternal copies of a sqh-gap43:mCherry transgene to label cell membranes [144] (Fig. 4.3a-d). To visualize myosin II, one maternal copy of a sqh-sqh:mCherry transgene (a fusion between mCherry and myosin regulatory light chain ubiquitously expressed under sqh promoter) was integrated into the fly.

4.2.2.2 Time-lapse imaging & tissue elongation measurement

The sample preparation, setup, and the objective choice are as described in Chapter 2, Section 2.2 Materials and methods. Instead of the standard area detectors, the Zeiss AiryScan detector was used to conduct faster and more sensitive imaging. Z-stacks were acquired at 1-μm steps and 1-min time intervals. Maximum intensity z-projections of 3 slices at different depths along the apical-basal axis were analyzed (Fig. 4.2b).

For each depth, tissue elongation measurements are as described in Chapter 2, Section 2.2 Materials and methods, except that $t = 0$ was the onset of tissue elongation at the most apical side of the tissue.

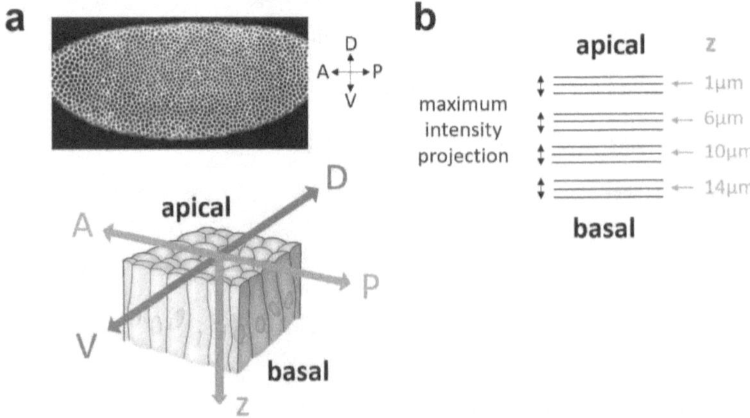

Figure 4.2. Orientations and image processing in the *Drosophila* germband epithelial tissue.
(a) Top: Confocal image of the lateral view the *Drosophila* embryo. Anterior (A), left; posterior (P),

right; dorsal (D), top; ventral (V), bottom. Bottom: Schematic of the 3-D structure of the germband epithelium. AP axis (blue) and DV axis (red) are perpendicular to the apical-basal axis (z, orange). z = 0 at the most apical plane and increases when it goes deeper towards the basal direction. (b) Analyses are done at different z-positions (z = 1, 6, 10, 14 μm) from the apical side to the basolateral side. Maximum intensity projection was done among images at a specific z and the two adjacent layers which are 1 μm away from the image along the z-axis.

4.2.2.3 3-D structure analysis

Each single z-slice of the confocal imaging stack was imaged in the plane of AP and DV axes of the tissue (Fig. 4.2 a, top). Anterior (A) was to the left. Posterior (P) was to the right. Dorsal (D) was to the top. Ventral (V) was to the bottom (Fig. 4.2 a, top). The z-stack was taken along the apical-basal axis with the most apical z-slice set as z = 0 (Fig. 4.2 a, bottom).

Slices at z = 1, 6, 10, and 14 μm were analyzed by doing maximum intensity projections with the adjacent slices above and below (Fig. 4.2b).

4.2.3 Results

4.2.3.1 *Drosophila* germband extension is shifted along the apical-basal axis

During *Drosophila* germband extension, the apical planar polarized actomyosin cortices of neighboring cells are tightly adhered to each other by adherens junctions (Fig. 4.1d) [96], which coordinates and regulates the collective movements of cells on their apical surface [40, 103, 104, 126-129]. Interestingly, there is much less actomyosin in the basolateral cortex of these cells, this actomyosin does not appear planar polarized, and there is no adhesive belt formed between the apical and basolateral cortex to organize or transmit forces from the apical cytoskeleton system to the basolateral cortex [96]. Therefore, the basolateral cellular and tissue-level behaviors might be distinct from the apical ones. To test this hypothesis, we analyzed both tissue and cellular mechanical behaviors by confocal imaging of the germband tissue at different z-positions along the apical-basal axis (Fig. 4.3 a-d).

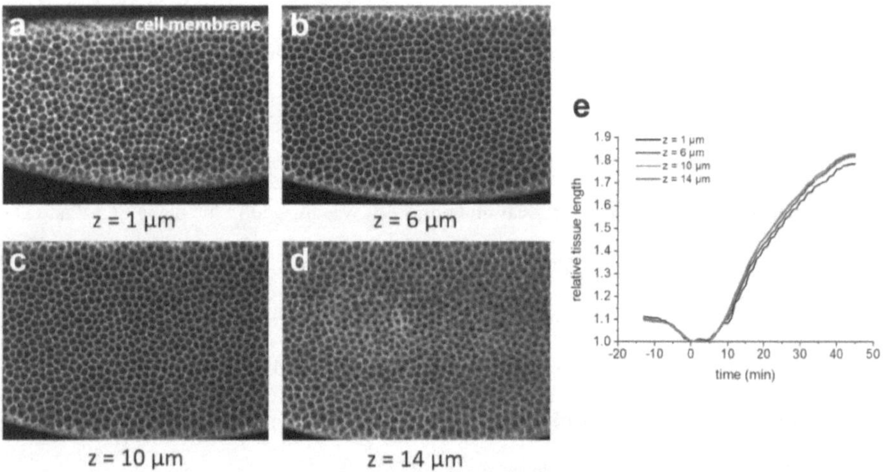

Figure 4.3. Tissue elongation in the *Drosophila* germband is shifted along the apical-basal axis. (a)-(d) Confocal images of the germband epithelial tissue at different z-positions along the apical-basal axis at the same time point from the same time-lapse movie. (e) Change of the relative tissue length over time at different z-positions along the apical-basal axis.

When we first looked at the tissue-level elongation, we found a 2-minute shift of the germband tissue movements between its apical and its basolateral parts. Surprisingly, the apical part of the tissue, which has the most cytoskeletal forces to drive movements and tissue flow, begins to elongate *after* the basolateral part of the tissue (Fig. 4.3e). This is consistent with recent studies showing the unexpected result that basolateral rosette formation during cell rearrangement precedes apical rosette formation [107] (Fig. 4.1c). These results indicate that enhanced protein localization and larger cell-generated forces do not necessarily simply correlate with cell movements or tissue deformations. Interestingly, this tissue elongation delay does not exist when comparing between basolateral parts of the tissue, as the relative tissue length lines at z = 6, 10, and 14 overlap with each other (Fig. 4.3e), suggesting that the apical part and the basolateral part of the germband tissue might utilize different mechanisms to elongate tissues, but that different basolateral regions of the germband tissue might share the same mechanism.

4.2.3.2 Cell patterns change along the apical-basal axis in the germband tissue

As proposed and discussed in Chapter 2, the fluidity of the germband epithelial tissue can be predicted by cell patterns. Next, we explored if cell patterns quantified by cell shape index p and cell shape alignment Q (details in Chapter 2 and Appendix A) change from the apical to the basolateral part of the tissue.

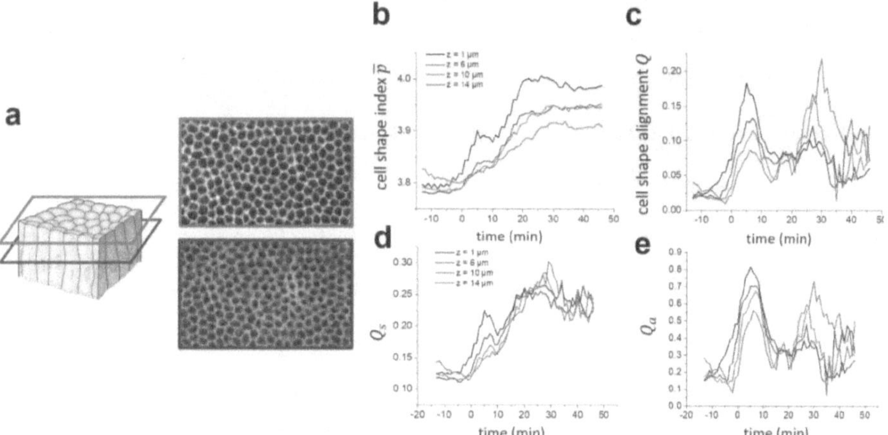

Figure 4.4. Cell patterns change along the apical-basal axis in the *Drosophila* germband tissue. (a) Left, schematic showing positions of cross-section views at z = 1 µm (red) and z = 14 µm (blue). Right, confocal images of cross-section views of the germband tissue at z = 1 µm (red) and z = 14 µm (blue). (b) Cell shape index p increases with time during axis elongation at both apical and basolateral parts of the germband tissue. Along the apical-basal axis, p decreases from the apical to the basolateral part. (c) Cell shape alignment Q first increases, then decreases, and then increases again with time during axis elongation, which peaks twice at around t = 2 and 32 min at both apical and basolateral parts of the germband tissue. During the first peak at t = 2 min, along the apical-basal axis, Q decreases from the apical to the basolateral part; at the second peak

at $t = 32$ min, Q increases from the apical to the basolateral part. (d) Q_s represents the contribution of cellular shape anisotropy to Q, while (e) Q_a represents the contribution of cell alignment to Q (details in Chapter 2 and Appendix A). At $t = 2$ min, both Q_s (d) and Q_a (e) show the same gradient as Q along the apical-basal axis indicating that both cell shape anisotropy and alignment contribute to this relationship; At $t = 32$ min, only Q_a (e) shows the same gradient as Q suggesting that cell alignment is the main contributor.

Germband epithelial cells in *Drosophila* embryos take on nearly isotropic hexagonal shapes at 10 minutes before body axis elongation when the tissue is static, but they change shape dramatically after body axis elongation starts. Consistent with this, the cell shape index p increases during tissue remodeling (Fig. 4.4b) while the cell shape alignment Q first increases, then decreases and then increases again showing two peaks at around $t = 2$ and 32 min (Fig. 4.4c). The overall qualitative changes of p and Q with time remains the same from the apical to basolateral part of the germband tissue (Fig. 4.4b and c).

We first examined if there is a difference in the value of p along the apical-basal axis. Interestingly, we found a gradient of p along the apical-basal axis: p decreases from the apical to the basolateral part as the depth z increases (Fig. 4.4b). We observe a similar gradient in Q at its first peak around $t = 2$ min, where Q decreases when the depth z increases (Fig. 4.4c). Cell shape alignment $Q = Q_s Q_a$ combines information about both cell shape anisotropy Q_s and cell alignment Q_a (See Appendix A.1.1). A further analysis shows that both Q_s and Q_a share the same gradient as Q along the apical-basal axis (Fig. 4.4d and e) indicating that both cell shape anisotropy and alignment contribute to the gradient of Q. Surprisingly, at the second peak of Q, the gradient of Q along the apical-basal axis reverses where Q increases with z while the gradient of p stays the same (Fig. 4.4b and c). At this time, only Q_a shows the same relationship as Q (Fig. 4.4 d and e) suggesting that cell alignment is the main contributor of Q here. These findings indicate that the germband epithelial tissue might use different mechanisms to tune its mechanical behaviors at the beginning and the end of the body axis elongation.

4.2.3.3 A predicted fluid-solid transition along the apical-basal axis in the germband epithelium

To explore if there is a tissue solid-fluid transition in the spatial dimension, in addition to the one observed in the temporal dimension (Chapter 2), we used methods developed in Chapter 2 to analyze fluidity at different z-positions along the apical-basal axis by combining analyses of the cell shape index p, cell shape alignment Q, and vertex coordination number z (detailed description and calculation in Chapter 2 and Appendix A). If we set the y axis as $\overline{p}_{corr}$, the average cell shape index corrected by z (described in Chapter 2, Section 2.3.5) and the x axis as the cell shape alignment Q (Fig. 4.5), the predicted solid-fluid transition line (Fig. 4.5, solid green lines) is $\overline{p}_{corr} = 3.818 + 4bQ^2$, where $b = 0.43$. If the tissue sits above the line, it is predicted to be more fluid-like, while the tissue behaves more solid-like if it is below the line.

Before and in the first 20 minutes of the germband extension, tissue mechanical behavior quantified at different z-positions transitions from solid-like to fluid-like behaviors over time (Fig. 4.5a). Interestingly, we also observe a gradient in tissue fluidity when we analyze from the most apical part to the basolateral part of the tissue. More specifically, our preliminary results show that at any given time, the tissue appears to become more solid-like when the depth z increases (Fig. 4.5a). This could be related to the two different types of gradients we observed in Q. At the first peak of Q, there are smaller forces driving tissue flow, so cells in the more solid-like basolateral part might still be able to resist forces and change shape less resulting in smaller p and Q (Fig. 4.4b and c). At the second peak of Q, the driving force is big enough to drive the whole tissue flow through cell rearrangement and cell shape change. The more solid-like basolateral part could have less cell rearrangements, so cells might have to change their shapes in this part to elongate the tissue. Therefore, cross-sectional cell shapes in the basolateral part are more aligned with each other than those in the apical part associated with higher Q around the second peak (Fig. 4.4c and e).

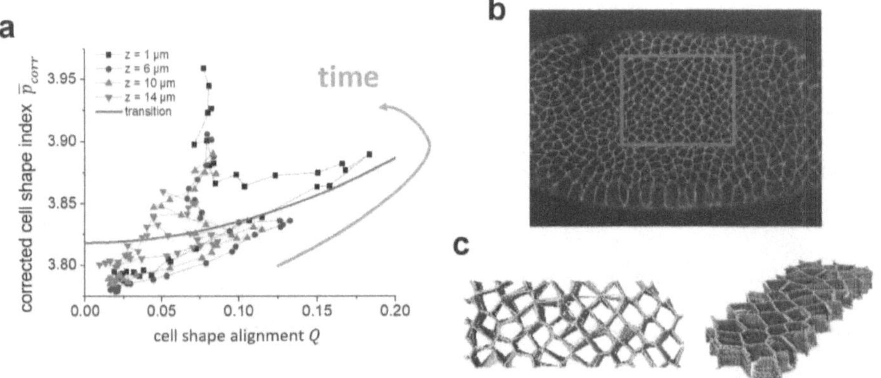

Figure 4.5. Tissue fluidity changes along the apical-basal axis in the germband epithelial tissue. (a) The cell shape alignment index Q and average corrected cell shape index $\overline{p}_{corr}$ for the germband tissue before and during axis elongation at different z-positions along the apical-basal axis. $\overline{p}_{corr}$ and Q were calculated for each time point and plotted along the time line (blue arrow) until t = 20 min. (n = 1 embryo with an average of about 300 cells analyzed per time point). The solid green line denotes the solid–fluid transition line in the anisotropic vertex model [72]. (b) Confocal image of the germband tissue analyzed (green area). (c) Reconstruction of the germband tissue by Para View showing the apical-basal twisting (left) and the high depth along the apical-basal axis (right).

4.2.4 Discussion and future work

In this ongoing work, I explore how tissue elongation, cell patterns, and tissue fluidity change along the apical-basal axis in the *Drosophila* germband epithelial tissue by combining high-depth live imaging and quantitative image analysis, which expands current research on mechanical behaviors of epithelial tissues from 2-D to 3-D. Preliminary results show that cell patterns, quantified by p and Q, change in gradients from the apical to basolateral side of the tissue, indicating potentially different mechanisms of controlling cellular behaviors along the apical-basal axis. The gradient changes with time, suggesting that these mechanisms change in the temporal dimension as well. Furthermore, we found that there appears to be a tissue solid-fluid transition along the apical-basal axis. In particular, the basolateral side of the tissue is predicted to be more solid-like compared with the apical side, which helps explain the gradients observed in cell patterns along the apical-basal axis.

Next, molecular mechanisms underlying these variations along the apical-basal axis need to be explored and tissue properties should be measured by quantifying cell rearrangements and directly measuring tissue mechanical properties. This work also provides necessary inputs to expand current 2-D models of epithelial tissues to 3-D. Currently we are adding 3-D mechanical properties into 3-D models and analyzing the influence of 3-D embryo curvatures on mechanical behaviors of the germband tissue (Fig. 4.5b and c).

4.2.5 Acknowledgement

For this ongoing project, I thank Professor Karen E. Kasza, Ph.D., for her supervision; Christian M. Cupo for assistance with data processing; Sassan Ostvar, Ph.D., for computational modeling; Dene Farrell and Jennifer Zallen, Ph.D., for the use of SEGGA, a segmentation and quantitative image analysis toolset; Adam Martin, Ph.D., for the sqh-gap43:mCherry fly stock, Jennifer Zallen, Ph.D., for the sqh-sqh:mCherry fly stock.

This work was supported by the National Science Foundation (NSF) Civil, Mechanical, and Manufacturing Innovation Grant 1751841 to Karen E. Kasza; K.E.K. holds a Burroughs Wellcome Fund Career Award at the Scientific Interface, Clare Boothe Luce Professorship, NSF CAREER Award, and Packard Fellowship.

4.3 Probing mechanics in the round window membrane of the inner ear

4.3.1 Introduction

The ear is the most important organ for hearing and balance in animals, including humans. The ear consists of three parts: the outer, middle, and inner ear (Fig. 4.6a). The inner ear, also known as the cochlea, transduces acoustic signals to electrical impulses that are transmitted via the vestibulocochlear nerve to the brain [200]. There are over 500 million people worldwide suffering from auditory and vestibular dysfunctions [110, 111]. Many of these diseases manifest themselves within the inner ear, such as Meniere's disease and sensorineural hearing loss. A longstanding clinical goal is to reliably and precisely deliver therapeutics into the inner ear to treat the auditory and vestibular dysfunctions. However, getting access to the inner ear is challenging because it is a space fully enclosed within the temporal bone of the skull, except for two membrane-covered portals connecting it to the middle ear space – the oval window and the round window (Fig. 4.6a) [201]. The oval window membrane is covered by the osseous footplate of the stapes, but not the round window membrane [201].

The round window membrane (RWM) is a biological membrane that separates the fluid-filled inner ear from the air-filled middle ear (Fig. 4.6). It plays an important role in modulating perilymphatic pressure and protecting the inner ear [201]. While the integrity of the RWM is essential for normal hearing, it is also the only portal into the inner ear without requiring bone perforation [110, 111]. The current method to treat inner ear diseases calls for injection of a therapeutic substance into the middle ear space, after which a portion of the substance diffuses across the RWM into the inner ear. The efficacy of this technique is limited by an inconsistent rate of molecular transport across the RWM. To access the inner ear directly for patient sampling and drug delivery, collaborators in Jeffrey Kysar's Group proposed to study mechanical properties in

the RWM and develop a novel approach by using ultra-sharp microneedles to perforate the RWM
[111]. This is among the very first work to explore mechanics in the RWM and might motivate
new clinical care standards for treating inner-ear diseases, which has the potential to save millions
of people from auditory and vestibular dysfunctions.

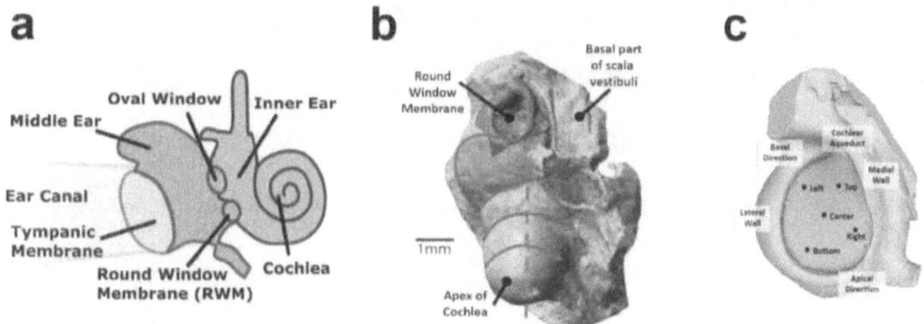

Figure 4.6. Anatomy of the ear. (a) Schematic of the ear. (b) Cochlea of guinea pig obtained through μ-CT. Image from (Watanabe *et al.*, 2014) [179]. Arrows represent turns of the spiral-shaped cochlear cavity from the apex to the RWM and dashed line represents the axis of the cochlea. (c) RWM (rose) connects to the round window bone structure (brown) at the sulcus.

In this collaboration, I leveraged the confocal imaging expertise developed in this book research to meet the different requirements for studies of the RWM. In probing the RWM structures and mechanical properties, my imaging experiments and results contributed to the reconstruction of the membrane geometry and fiber orientations. We are currently studying the responses of the RWM to deformations by customizing the imaging platform to control the pressure underlying the RWM. To develop the new method to deliver drugs into the inner ear, the Kysar Group fabricated a 3-D printed ultra-sharp needle to protrude the membrane, generating a recoverable hole to let the drugs pass through. My imaging results were used for analyses of the shape of the hole made by the needle, the change of tissue mechanics around the hole, and the recovery of the tissue following this perturbation using confocal imaging. These interdisciplinary and collaborative projects extend my methods developed in this book for the *Drosophila* germband epithelial tissue to other types of tissue, showing the potential wider applications of the research in this book.

4.3.2 Materials and methods

4.3.2.1 Harvesting guinea pig cochleae

The guinea pig cochleae were harvested by collaborators in Jeffrey Kysar's Group and Anil Lalwani's Group at Columbia University and Columbia University Medical Center.

Carcasses of mature guinea pigs (Hartley, Charles River, Massachusetts) with no history of middle ear disease were obtained via tissue sharing facilitated by the Institute of Comparative Medicine at Columbia University Medical Center. All animals were euthanized using pentobarbital overdose for the purpose of harvesting their trachea. Immediately following euthanasia, the intact temporal bone of the guinea pig was harvested using blunt dissection. A drill was used to remove the surrounding bone, exposing a clear, wide-angle view of the RWM.

To analyze mechanics of the membrane, the RWM remained attached to the sulcus of the bone for all subsequent processing and imaging steps. The resulting specimen was rinsed in 0.9% phosphate buffered saline (PBS - Sigma Aldrich D5652) and inspected for gross membrane perforations and fractures of the RWM sulcus and niche. The RWM was then either fixed overnight in 10% buffered formalin or stored in PBS and refrigerated.

To analyze mechanics of the microneedle-perforated membrane, the resulting specimen was rinsed with 0.9% saline solution and inspected for gross membrane perforations and fractures of the RWM niche. If perforation of the RWM could not be performed immediately, the specimen was refrigerated in 0.9% saline solution (up to a maximum of 24 hours) prior to further experimental use. During perforation experiments, small amounts of sterile 0.9% saline solution were applied at regular intervals to keep the membrane from drying.

4.3.2.2 Confocal microscopy

Imaging for analyzing membrane mechanics. Rhodamine B has been used by other researchers to stain elastic fibers [180]. The RWM is first immersed in a 1mM solution of Rhodamine B in PBS overnight and then rinsed with PBS and immersed in 100 mL of PBS 1h before imaging. The RWM is then rinsed a final time with PBS and placed in a MatTek glass bottom dish (No. 1.5). Imaging was done on an inverted confocal laser scanning microscope Zeiss LSM 880, Axio Observer with a 10× objective (EC Plan-Neofluar 10×/0.30 M27) or a 20× objective (Plan-Apochromat 20×/0.8 M27). A stack of images was generated at several focal heights spaced 1 μm and 5 μm apart for the 20X objective and the 10X objective, respectively, to obtain 3D images of the RWM. These images were then projected in the stacking direction (maximum intensity z-projection) to obtain a global image with the visible fibers.

Imaging for analyzing perforated membranes. Prior to imaging, the RWM was immersed in a 1 mM solution of Rhodamine B in phosphate buffered saline (PBS) for 1 hour. It was then rinsed several times with PBS and placed in a MatTek glass bottom dish (No. 1.5). The imaging of the perforated membrane was done on an inverted confocal laser scanning microscope Zeiss LSM 880, Axio Observer with a 10× objective (EC Plan-Neofluar 10×/0.30 M27) or a 20× objective (Plan-Apochromat 20×/0.8 M27). An excitation wavelength of 561 nm was chosen for the laser, and emitted light from 576 nm to 682 nm was allowed to pass to the detector. A stack of images was generated at several focal heights spaced 1 μm and 5 μm apart for the 20× objective and the 10× objective, respectively. These images were then projected in the stacking direction (maximum intensity z-projection) to obtain a global image with the visible perforation.

4.3.3 Results

4.3.3.1 RWM membrane curvature and fiber alignment analysis by confocal imaging

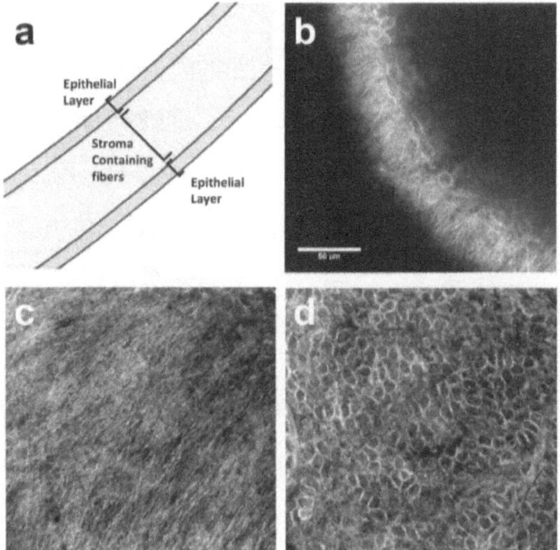

Figure 4.7. Characterization of the connective tissue in the RWM. (a) Schematic of structures shown in a single slice of confocal image stacks. (b) Confocal image of a single slice of image stack. Both epithelial layers and connective tissue layer visible. (c) Z-projected confocal image of connective tissue layer. (d) Z-projected confocal image of outer epithelial layer.

To study mechanics in the RWM, first we need to understand its structure. The RWM has three layers: the fiber-containing stroma layer, which is covered by an epithelial layer on each side (Fig. 4.7a and b). To obtain a 2-D image to process fiber directionality based upon 20X Rhodamine B imaging, we needed to flatten the stack from 3-D to 2-D. The membrane is very thin, so each slice of the 3-D stack contains only a very small strip of membrane (Fig. 4.7b) with both the epithelial layers and the connective tissue layer visible. Our collaborators developed a numerical

flattening method, which successfully split the connective tissue layer (Fig. 4.7c) and the epithelial

layers (Fig. 4.7d).

Because the RWM was harvested, processed, and imaged with temporal bones, the *in vivo*

geometry of the RWM was reserved. With the 3-D imaging stacks obtained by Rhodamine B

staining and confocal microscopy, we successfully reconstructed the geometry of the RWM (Fig.

4.8a-c), which was used by our collaborators to analyze the RWM curvatures. Meanwhile, the

fiber orientations are clearly shown in the confocal images (Fig. 4.8d), which was used to analyze

fiber alignment.

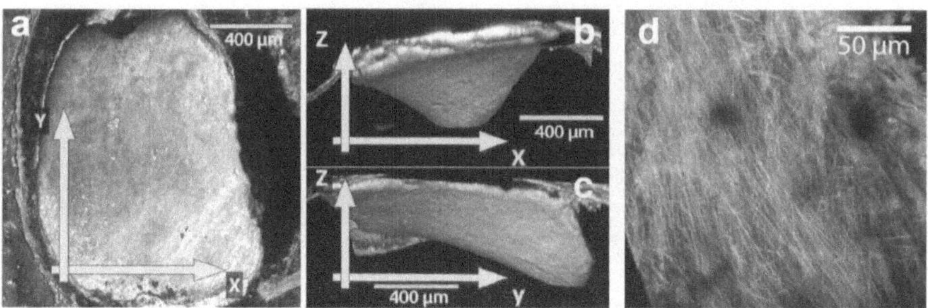

**Figure 4.8. Reconstruction of the RWM and analysis of the RWM curvature and fiber
alignment.** (a)-(c) Geometry of the RWM reconstructed from confocal image stacks. (a) Top view
from z-axis as seen from the middle ear. Bottom region (smaller y-values) corresponds to the
flatter region of the membrane while the top region (larger y-values) corresponds to a deeper and
more conically shaped region. (b) Side view from y-axis. Conical shape of membrane is evident.
(c) Side view from x-axis. Transition from flat to conical shape is seen as y-value increases. (d)
Confocal image of both collagen and elastic fibers (yellow) stained by Rhodamine B.

4.3.3.2 Perforation of the RWM via direct 3-D printed microneedles

Besides analyzing the RWM mechanical structures such as the curvatures and fiber

alignments, our experimental platform was also successfully used to analyze mechanics in the

RWM after perforation via 3-D printed ultra-sharp microneedles (Fig. 4.9a). These results are being used to develop a new method to deliver drugs into the inner ear by making a recoverable hole in the RWM.

Using our imaging platform, the shape and area of the perforations created with the microneedles were accurately recorded (Fig. 4.9b). These results were used by our collaborators to analyze mechanics in the membrane after the perforation. We found that the perforation is lens-shaped and remains partially open after the microneedle is removed. By analyzing the size of the perforations from different needle designs and timelines, for example, we can optimize the needle design and recovery protocol after the procedure conducted in the inner ear. In addition, imaging with higher magnification reveals the fiber orientations in the membrane (Fig. 4.9c), which allowed our collaborators to analyze detailed mechanical conditions in perforated membranes and optimize their needle and procedure design.

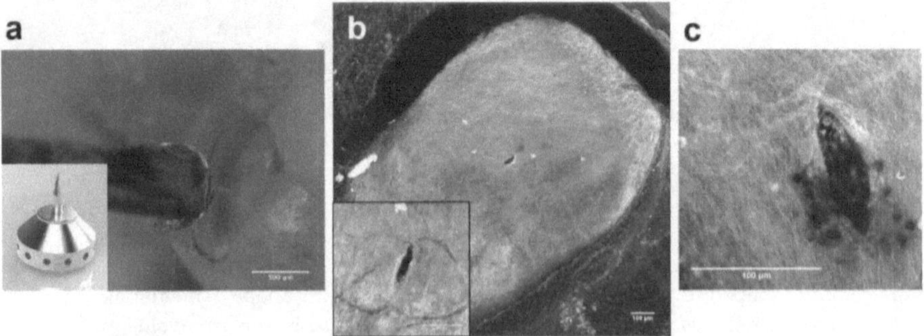

Figure 4.9. Perforation of the RWM via direct 3-D printed microneedles. Images reprinted by permission from Biomedical Microdevices. (a) Image of a microneedle attached to the microindentor perforating the RWM of a guinea pig. Inset, 3-D model of the microneedle. (b) Confocal image of a whole RWM with a perforation made by a microneedle in the center. Inset, zoomed-in confocal image of a perforation made by a microneedle in a RWM. Scale bar, 100 μm. (c) Confocal image of a RWM zoomed in around a perforation made by a microneedle with fibers shown. The major axis of the lens-shaped perforation is aligned with the fibers of connective

tissue and fiber reorientation at the crack tip can be observed. The small dark circular features are cellular debris that is occluding the view.

4.3.4 Discussion and future work

In this section, interdisciplinary and collaborative projects related to the RWM are described, expanding my methods developed in this book from *Drosophila* germband epithelial tissue to other types of tissue and from *in vivo* work to *ex vivo / in vitro* work. By customizing my experimental platform, studies on different tissues under complex mechanical conditions can be conducted. For example, we are currently further analyzing the deformation response of the RWM by changing pressures underlying the membrane, mimicking the perilymphatic pressure change in the inner ear. A customized imaging platform (Fig. 4.10) was designed, which makes use of live imaging of the sample while the pressure underneath changes.

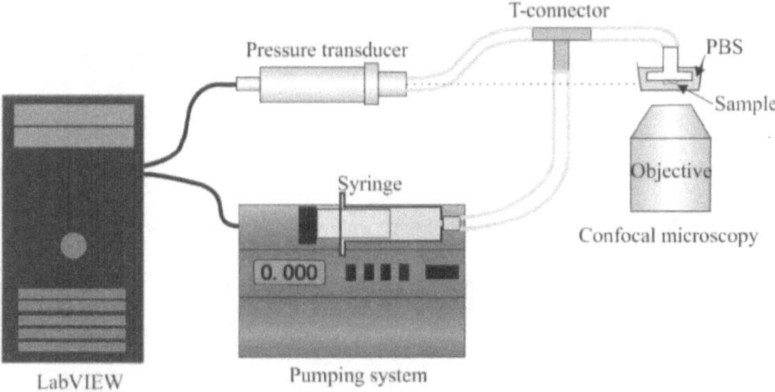

Figure 4.10. Analyze deformation response of the RWM. Schematic of a customized imaging platform to image samples with different deformations induced by changing pressures.

Chapter 5. Conclusion and Future Directions

5.1 Conclusion

5.1.1 Motivating questions

The work in this book was motivated by a lack of fundamental understanding of the roles of mechanics in morphogenesis. An improved understanding of the mechanical aspects of morphogenesis will be essential in filling the gap in our knowledge in connecting molecular-scale activities to larger-scale tissue behaviors in living organisms. This book explores mechanical factors in epithelial tissue morphogenesis during embryonic development of *Drosophila* by developing and using a systematic, quantitative, *in vivo* experimental approach to address important open questions in the field of the mechanics of morphogenesis.

First, this book investigated what mechanical factors are involved in the morphogenesis of epithelial tissues. More specifically, we explored the roles of epithelial tissue mechanical properties in morphogenesis *in vivo*. Biological tissues behave as living materials to dramatically flow and reorganize to form functional tissues and organs. Recent experimental and computational studies both highlight that these morphogenetic events can be linked with a special tissue mechanical property – tissue fluidity [18, 53, 64, 105, 118, 119]. Yet, tissue fluidity in anisotropic epithelial tissues such as the *Drosophila* germband tissue, remains poorly understood. To answer these questions, in Chapter 2, we combined confocal imaging and quantitative image analysis with a vertex model developed by our collaborators to study the germband epithelium during *Drosophila* body axis elongation.

Second, this book explored how cells control these mechanical factors to tune tissue-level behaviors during epithelial tissue morphogenesis. In particular, we investigated how the properties of epithelial cells are regulated to tune the fluidity of the tissues in which they reside and revealed the role of cell-cell adhesion in epithelial tissue mechanics. Unlike contractile tension generated by myosin II, cell-cell adhesion mediated by E-cadherin plays interesting 'dual roles'. It physically adheres cells to each other to make the tissue more solid-like, but also organizes and transmits forces that promote tissue remodeling to make it more fluid-like. It is unclear how cell-cell adhesion enacts its dual functions in epithelial tissue mechanics and morphogenesis. To explore the role of cell-cell adhesion in controlling epithelial tissue mechanics, in Chapter 3, we combine genetic approaches, live confocal imaging and quantitative image analysis to study the effects of cell-cell adhesion levels on cellular and tissue behaviors.

Third, this book studied how mechanics of epithelial tissues transitions from 2-D to 3-D. In most studies, epithelial tissues are simplified into 2-D structures by only considering the apical side [7, 40-42 67-72], partially due to a lack of live imaging methods to take high-depth images and of a non-invasive platform to probe mechanics *in vivo*. The importance of 3-D mechanics in tissue mechanics has been proven in recent *in vivo* studies [106-109], however, little is known about the 3-D mechanical behaviors in the *Drosophila* epithelium. To address this question, in Chapter 4, we expanded our experimental approach from Chapters 2 & 3 to study mechanical behaviors of the *Drosophila* germband epithelial tissue from the apical side to the basolateral side to have a more complete image of mechanics of the germband tissue in 3-D.

Finally, in Chapter 4, this book also included work that applies the experimental approaches described above to explore mechanical behaviors in other types of biological materials and develop clinical applications.

5.1.2 Work performed

To address the questions above, we explored the roles of mechanics during epithelial tissue morphogenesis *in vivo*, which can be quite challenging. On the one hand, in living tissues, mechanical factors and biological factors are strongly coupled with each other and actively respond to the environment. On the other hand, the field lacks experimental methods to study the mechanics of epithelial tissues *in vivo*, especially non-invasive methods that minimally affect tissue behaviors. In this book, we developed a systematic, quantitative, *in vivo* experimental platform, which can be customized to investigate different mechanical factors during epithelial tissue morphogenesis and was successfully combined with theoretical modeling and used in other types of biological materials. Here is a summary of the work performed.

To explore the tissue fluidity during epithelial tissue morphogenesis *in vivo*, in Chapter 2, we used high-resolution confocal fluorescence imaging to take time-lapse movies of embryonic development in the *Drosophila* embryo and analyzed the shape and packings of cells in tissues. We combined our quantification of cell rearrangements and cell patterns, especially cell shape index p and cell shape alignment Q, with our collaborators' vertex model of anisotropic epithelial tissues to generate a fit-parameter-free prediction of the solid-fluid transition in the developing germband epithelial tissue. To further test our results, genetics mutants were used to modify mechanical conditions in the tissue and their tissue fluidity was studied.

To investigate the role of cell-cell adhesion between epithelial cells in tuning epithelial tissue mechanics and morphogenesis, in Chapter 3, we used genetic approaches to systematically modulate cell-cell adhesion levels in *Drosophila* embryonic epithelial tissues, and combined live confocal fluorescence imaging and quantitative image analysis to study the effects of cell-cell adhesion levels on cellular and tissue behaviors. The cell-cell adhesion level is quantified by the

E-cadherin protein level, which was measured from fixed samples both manually and automatically. The influence of cell-cell adhesion levels on protein dynamics was quantified by fluorescence recovery after photobleaching (FRAP). To probe how cell-cell adhesion levels affect the cytoskeletal system, genetic approaches were used to tune E-cadherin levels with myosin II, F-actin, and moesin fluorescently tagged, whose intensities and polarities were quantified both manually and automatically.

To further improve the capacities of our experimental platform, we expanded our work from 2-D to 3-D and applied it to other tissue types for clinical application development. First, we optimized the confocal microscope setup to take high-depth fluorescence live imaging of the *Drosophila* germband tissue. The cell pattern and tissue change quantification methods and the tissue fluidity prediction methods were combined with these 3-D images to explore how mechanical behaviors change along the apical-basal axis. Second, we switched our imaging platform from imaging *Drosophila* epithelial tissues fluorescently tagged by transgenes to imaging the guinea pig round window membrane fluorescently dyed by Rhodamine B [180]. Third, we customized our imaging platform connecting it with a controlled pumping system to image round window membranes under different deformations by pressure change.

5.1.3 Significant findings

In this book, we conducted a series of experimental studies to explore mechanical properties of the *Drosophila* germband epithelial tissue and the molecular mechanisms underlying these tissue mechanical behaviors, which are vital and practical in pursuing the primary scientific objective of better understanding the role of mechanics in epithelial tissue morphogenesis. Here we summarize and discuss significant results from each of these studies.

5.1.3.1 Epithelial tissue fluidity

The most practical and appealing finding from Chapter 2 is providing an approach to read out epithelial tissue mechanical behaviors simply by looking at a snapshot of cell shapes in the tissue, with no fit parameters. In this work, we show that cell shape, cell alignment, and packing disorder can be used to understand and predict whether an anisotropic tissue flows and remodels like a fluid or maintains its shape like a solid. We found that during convergent extension, the *Drosophila* germband might be viewed as a transition to more fluid-like behavior to help accommodate dramatic tissue flows, which raises the possibility that the mechanical properties of developing tissues might be tuned to become more fluid-like during rapid morphogenetic events.

This approach is independent of the underlying origin of anisotropy, and therefore can be used to predict mechanical behavior of tissues even when external and internal stresses cannot be directly measured. Meanwhile, cell shape patterns are easy to access experimentally from snapshots of cell packings in tissues, even in systems where time-lapse live imaging of cell rearrangement and tissue flow is not possible.

This novel method to understand mechanics that controls tissue flow has the potential to be applied in other tissue types and model organisms. Also, it offers a powerful tool for engineers, physicists and biologists to study complex tissue behaviors in a broad range of morphogenetic processes occurring in developing embryos *in vivo* or organoid systems *in vitro* and probe mechanics in tissues where direct mechanical measurements are currently impossible, such as human embryonic tissues.

5.1.3.2 Cell-cell adhesion in epithelial tissue mechanics

The work in Chapter 3 provides a fundamental understanding of the "dual roles" of cell-cell adhesion in epithelial tissue mechanics and how cell-cell adhesion influences cellular behaviors and tissue fluidity. The most surprising findings in this study are the biphasic dependencies of cell rearrangements, cell shape patterns, and tissue fluidity on cell-cell adhesion levels, which are linked to each other by cell shape patterns. In particular, tissues comprising cells with both lower or higher cell-cell adhesion levels tend to rearrange faster and show cell patterns indicating more fluid-like tissue behaviors. Further studies of behaviors of molecules coupled with cell-cell adhesion show that cell-cell adhesion levels also influence other molecules, such as junctional myosin II, and tune junctional dynamics, such as cell-cell contact contraction and growth, protein dynamics, which helps to explain the influence of cell-cell adhesion on cellular behaviors and tissue fluidity in epithelial tissues.

This work advances our understanding of mechanics of epithelial tissue morphogenesis and provides quantitative experimental approaches to explore tissue mechanics *in vivo* and at multi-scales. The methods developed in this study can be extended and customized to explore other molecules in other tissue types. The findings suggest that cell-cell adhesion plays active dual roles in modulating epithelial tissue remodeling through direct and indirect influences, which helps explain how confluent epithelial tissues remodel while maintaining their integrity. This fundamental research provides necessary inputs for building more accurate models of epithelial tissues by considering the complex role of cell-cell adhesion and might shed light on the improper regulation of tissue mechanics associated with birth defects, errant wound healing, and cancer metastasis.

5.1.3.3 Tissue mechanics in 3-D and in the RWM

In Chapter 4, the experimental platform developed for exploring mechanics in 2-D in the *Drosophila* germband epithelial tissue is expanded to be used in 3-D and other tissue types.

When analyzing *Drosophila* germband tissue mechanical behaviors in 3-D, the most interesting preliminary result is that, unlike the biphasic dependencies in Chapter 3, cell shape patterns show a one-way gradient from the apical to the basolateral part of the tissue, indicating there are different mechanisms controlling cellular behaviors at different parts of the tissue along the apical-basal axis. Furthermore, there appears to be a tissue fluidity transition along the apical-basal axis, where tissue domains in the basolateral side are predicted to behave more solid-like compared with tissues in the apical side. This ongoing project expands the current experimental studies to 3-D and provides insights into building 3-D computational models. It reveals that living tissues undergoing morphogenetic events are to be considered 3-D.

When adapting our experimental platform to investigation of mechanical properties of the round window membrane (RWM) in the guinea pig inner ear, we succeed in transferring methods developed for the *Drosophila* germband epithelial tissue to membrane tissues in guinea pigs. Super high-resolution images of the RWM were acquired from our system to analyze geometrical details and fiber orientations of the RWM by our collaborators. Furthermore, my platform helped develop a new method to deliver drugs into the inner ear by using a 3-D printed ultra-sharp needle fabricated by my collaborators to perforate the membrane to generate a recoverable hole to let drugs pass through. Super-high-resolution images of the perforated RWM were generated for our collaborators to analyze the shape of the hole made by the needle, the change of fibers around the hole, and the recovery of the tissue following this perturbation. Our collaborative work shows that the RWM has a complex curved geometry that improves its function compared to a flat membrane, and establishes a foundation for using microneedles to controllably perforate membranes for medical applications, setting an example of how to connect bench work to clinical applications.

5.2 Future directions

This book has addressed several important questions in the field of mechanics of epithelial tissue morphogenesis, but also provoked many new research directions and opportunities to move forward. In this section, we present and discuss several directions where future work can and should be done.

In the work from Chapter 2 to explore tissue fluidity during *Drosophila* epithelial tissue morphogenesis *in vivo*: (1) We did not investigate the underlying origin of mechanical anisotropy, so it will be interesting to explore experimentally how the nature of internal and external forces contribute to tissue mechanics, cell rearrangement, and tissue flows in the germband. (2) We focused on an area that is a few micrometers deep from the most apical part of the tissue and simplified the epithelial tissue from 3-D to 2-D, so future study of mechanics of the basolateral part of the germband tissue might give us a more complete understanding of epithelial tissue mechanics. Some preliminary work has been done in Chapter 4 and shows interesting results. (3) Incorporating these internal and external forces and 3-D features into current vertex models could give us more sophisticated models to understand the diverse behaviors of living tissues. (4) The approaches we developed here to read out tissue fluidity can be used in other tissues such as in other developing epithelial tissues and different organs *in vivo*, or organoid systems *in vitro*. A tissue fluidity map across different tissue types and animal species can be built using our approaches, and interesting results might be found by comparing these fluidity results from different tissues and animals.

In the work from Chapter 3 to investigate the role of cell-cell adhesion in tuning epithelial tissue mechanics and morphogenesis: (1) We mainly analyzed cell-cell adhesion and tissue behaviors at the most apical part of the tissue. However, the cell-cell adhesion-mediating protein E-cadherin is distributed around the whole cortex of epithelial cells and its clustering and coupling with the cytoskeleton at the basolateral side is different from that at the apical side. It would be interesting to explore the E-cadherin distribution at the basolateral part of the tissue and how it can be related to cellular and tissue mechanical behaviors there. (2) Current vertex models ignore details of cell-cell adhesion and contractile tension. Incorporating our study of the role of cell-cell adhesion into vertex models could provide more precise tools to understand and gain insight into tissue morphogenetic processes. (3) Current models for epithelial tissues such as vertex models usually lack the temporal dimension. Our experimental work provides timescale information such as the rate of cell rearrangement and the speed of tissue elongation. Future work incorporating timescales into vertex models could provide better understanding of the dynamics of tissue mechanical behaviors.

In the work from Chapter 4 to expand our current work: (1) Current 3-D models lack accurate information of local mechanical properties. Integrating our 3-D mechanical property results into 3-D models could potentially generate more accurate simulations of living tissues. (2) Recent work shows that 3-D geometric constraints are related to the modulation of cell rearrangement. It will be interesting to include 3-D mechanical properties in analyzing the influence of embryo curvatures on mechanical behaviors of the germband tissue. (3) Currently, we are extending our understanding of mechanics of the RWM by analyzing the deformation response of the RWM, which could provide more information on RWM mechanical properties and mimic the perilymphatic pressure change in the inner ear.

5.3 Closure

The findings in this book advanced our understanding of *in vivo* mechanics of epithelial tissue morphogenesis and the methods developed in this book provided a practical, quantitative, and appealing platform to explore mechanics in living tissues during morphogenesis. Future work should further elucidate molecular mechanisms underlying epithelial tissue mechanical behaviors and extend the analysis to 3-D. More sophisticated computational models can be built based on these newly quantified *in vivo* tissue mechanics. Ultimately, a full understanding of mechanical factors during epithelial tissue morphogenesis will help fill the gap in our knowledge in connecting molecular-scale activities to larger-scale tissue behaviors, directly building tissues with desired shapes and structures in the lab, and shedding light on human diseases associated with improper regulation of tissue mechanics such as birth defects, aberrant wound healing, and cancer metastasis.

References

1. C. H. Waddington, *Principles of embryology*, First edition (George Allen & Unwin, 1956).

2. S. F. Gilbert, *Developmental biology*, Ninth edition (Sinauer Associates, 2003).

3. J. A. Davies, *Mechanisms of morphogenesis*, Second edition (Elsevier/AP, 2013).

4. D. W. Thompson, *On growth and form*, Second edition (University Press, 1959).

5. A. Šiber, P. Ziherl, *Cellular patterns* (CRC Press, Taylor & Francis Group, 2018).

6. D. J. Montell, Morphogenetic cell movements: diversity from modular mechanical properties. *Science* **322**, 1502–1505 (2008).

7. H. G. Lee, D. C. Zarnescu, B. MacIver, G. H. Thomas, The cell adhesion molecule roughest depends on heavy-spectrin during eye morphogenesis in *Drosophila*. *Journal of Cell Science* **123**, 277–285 (2010).

8. J. W. Spurlin, *et al.*, Mesenchymal proteases and tissue fluidity remodel the extracellular matrix during airway epithelial branching in the embryonic avian lung. *Development* **146**, dev175257 (2019).

9. D. Tzou, *et al.*, Morphogenesis and morphometric scaling of lung airway development follows phylogeny in chicken, quail, and duck embryos. *EvoDevo* **7**, 12 (2016).

10. K. Goodwin, *et al.*, Smooth muscle differentiation shapes domain branches during mouse lung development. *Development* **146**, dev181172 (2019).

11. J. F. Durel, N. L. Nerurkar, Mechanobiology of vertebrate gut morphogenesis. *Current Opinion in Genetics & Development* **63**, 45–52 (2020).

12. N. L. Nerurkar, C. Lee, L. Mahadevan, C. J. Tabin, Molecular control of macroscopic forces drives formation of the vertebrate hindgut. *Nature* **565**, 480–484 (2019).

13. K. E. Garcia, C. D. Kroenke, P. V. Bayly, Mechanics of cortical folding: stress, growth and stability. *Philosophical Transactions of the Royal Society B: Biological Sciences* **373**, 20170321 (2018).

14. A. Goriely, *et al.*, Mechanics of the brain: perspectives, challenges, and opportunities. *Biomechanics and Modeling in Mechanobiology* **14**, 931–965 (2015).

15. J. M. Fletcher, T. J. Brei, Introduction: Spina Bifida—A multidisciplinary perspective. *Development Disabilities Research Reviews* **16**, 1–5 (2010).

16. A. J. Copp, *et al.*, Spina bifida. *Nature Reviews Disease Primers* **1**, 1–18 (2015).

17. V. Ajeti, *et al.*, Wound healing coordinates actin architectures to regulate mechanical work. *Nature Physics* **15**, 696–705 (2019).

18. R. J. Tetley, *et al.*, Tissue fluidity promotes epithelial wound healing. *Nature Physics* **15**, 1195–1203 (2019).

19. T. Sugino, *et al.*, Morphological evidence for an invasion-independent metastasis pathway exists in multiple human cancers. *BMC Medicine* **2**, 9 (2004).

20. L. A. Tashireva, *et al.*, Single tumor cells with epithelial-like morphology are associated with breast cancer metastasis. *Frontiers Oncology* **10** (2020).

21. A. Akhtar, *et al.*, Neurodegenerative diseases and effective drug delivery: A review of challenges and novel therapeutics. *Journal of Controlled Release* **330**, 1152–1167 (2021).

22. C. M. Hall, E. Moeendarbary, G. K. Sheridan, Mechanobiology of the brain in ageing and Alzheimer's disease. *European Journal of Neuroscience* **00**, 1-28 (2020).

23. J. C. P. Coninck, *et al.*, Network properties of healthy and Alzheimer brains. *Physica A: Statistical Mechanics and its Applications* **547**, 124475 (2020).

24. M. P. Lutolf, J. A. Hubbell, Synthetic biomaterials as instructive extracellular microenvironments for morphogenesis in tissue engineering. *Nature Biotechnology* **23**, 47–55 (2005).

25. P. C. Bessa, M. Casal, R. L. Reis, Bone morphogenetic proteins in tissue engineering: the road from the laboratory to the clinic, part I (basic concepts). *Journal of Tissue Engineering and Regenerative Medicine* **2**, 1–13 (2008).

26. A.-L. Duchemin, H. Vignes, J. Vermot, R. Chow, Mechanotransduction in cardiovascular morphogenesis and tissue engineering. *Current Opinion in Genetics & Development* **57**, 106–116 (2019).

27. B. Hu, *et al.*, Orthogonally engineering matrix topography and rigidity to regulate multicellular morphology. *Advanced Materials* **26**, 5786–5793 (2014).

28. R. D. Kamm, *et al.*, Perspective: The promise of multi-cellular engineered living systems. *APL Bioengineering* **2**, 040901 (2018).

29. C. Gilbert, T. Ellis, Biological engineered living materials: growing functional materials with genetically programmable properties. *ACS Synthetic Biology* **8**, 1–15 (2019).

30. J. C. Serrano, S. K. Gupta, R. D. Kamm, M. Guo, In pursuit of designing multicellular engineered living systems: a fluid mechanical perspective. *Annual Review of Fluid Mechanics* **53**, 411–437 (2021).

31. B. Altmann, *et al.*, Differences in morphogenesis of 3D cultured primary human osteoblasts under static and microfluidic growth conditions. *Biomaterials* **35**, 3208–3219 (2014).

32. P. Samal, C. van Blitterswijk, R. Truckenmüller, S. Giselbrecht, Grow with the flow: when morphogenesis meets microfluidics. *Advanced Materials* **31**, 1805764 (2019).

33. E. A. Margolis, *et al.*, Stromal cell identity modulates vascular morphogenesis in a microvasculature-on-a-chip platform. *Lab on a Chip* **21**, 1150–1163 (2021).

34. Aristotle, *Aristotle: History of animals, Books I-III* (Harvard University Press, 1965).

35. C. Darwin, J. Huxley, *The Origin of Species: 150th Anniversary Edition*, 150th Anniversary ed. edition (Signet, 2003).

36. M. L. Coleman, M. F. Olson, Rho GTPase signaling pathways in the morphological changes associated with apoptosis. *Cell Death & Differentiation* **9**, 493–504 (2002).

37. S. Y. Moon, Y. Zheng, Rho GTPase-activating proteins in cell regulation. *Trends in Cell Biology* **13**, 13–22 (2003).

38. O. Pertz, Spatio-temporal Rho GTPase signaling – where are we now? *Journal of Cell Science* **123**, 1841–1850 (2010).

39. A. J. Ridley, Rho GTPase signalling in cell migration. *Current Opinion in Cell Biology* **36**, 103–112 (2015).

40. J. A. Zallen, E. Wieschaus, Patterned gene expression directs bipolar planar polarity in *Drosophila. Developmental Cell* **6**, 343–355 (2004).

41. K. D. Irvine, E. Wieschaus, Cell intercalation during *Drosophila* germband extension and its regulation by pair-rule segmentation genes. *Development* **120**, 827–841 (1994).

42. A. C. Paré, *et al.*, A positional Toll receptor code directs convergent extension in *Drosophila. Nature* **515**, 523–527 (2014).

43. C.-P. Heisenberg, Y. Bellaïche, Forces in tissue morphogenesis and patterning. *Cell* **153**, 948–962 (2013).

44. T. Lecuit, P.-F. Lenne, E. Munro, Force generation, transmission, and integration during cell and tissue morphogenesis. *Annual Review of Cell and Developmental Biology* **27**, 157–184 (2011).

45. R. M. Herrera-Perez, K. E. Kasza, Biophysical control of the cell rearrangements and cell shape changes that build epithelial tissues. *Current Opinion in Genetics & Development* **51**, 88–95 (2018).

46. F. Serwane, *et al.*, *In vivo* quantification of spatially varying mechanical properties in developing tissues. *Nature Methods* **14**, 181–186 (2017).

47. D. Kong, F. Wolf, J. Großhans, Forces directing germ-band extension in *Drosophila* embryos. *Mechanisms of Development* **144**, 11–22 (2017).

48. C. M. Nelson, *Tissue morphogenesis: methods and protocols* (Springer, 2015).

49. E. H. Chan, P. Chavadimane Shivakumar, R. Clément, E. Laugier, P.-F. Lenne, Patterned cortical tension mediated by N-cadherin controls cell geometric order in the *Drosophila* eye. *eLife* **6**, e22796 (2017).

50. S. Budday, P. Steinmann, E. Kuhl, The role of mechanics during brain development. *Journal of the Mechanics and Physics of Solids* **72**, 75–92 (2014).

51. C. D. Buckley, *et al.*, The minimal cadherin-catenin complex binds to actin filaments under force. *Science* **346** (2014).

52. S. de Beco, J.-B. Perney, S. Coscoy, F. Amblard, Mechanosensitive adaptation of E-cadherin turnover across adherens junctions. *PLOS ONE* **10**, e0128281 (2015).

53. K. V. Iyer, R. Piscitello-Gómez, J. Paijmans, F. Jülicher, S. Eaton, Epithelial viscoelasticity is regulated by mechanosensitive E-cadherin turnover. *Current Biology* **29**, 578-591.e5 (2019).

54. M. von Dassow, L. A. Davidson, Natural variation in embryo mechanics: gastrulation in *Xenopus laevis* is highly robust to variation in tissue stiffness. *Developmental Dynamics* **238**, 2–18 (2008).

55. K. Doubrovinski, M. Swan, O. Polyakov, E. F. Wieschaus, Measurement of cortical elasticity in *Drosophila melanogaster* embryos using ferrofluids. *PNAS* **114**, 1051–1056 (2017).

56. A. D'Angelo, K. Dierkes, C. Carolis, G. Salbreux, J. Solon, *In vivo* force application reveals a fast tissue softening and external friction increase during early embryogenesis. *Current Biology* **29**, 1564-1571.e6 (2019).

57. K. Bambardekar, R. Clément, O. Blanc, C. Chardès, P.-F. Lenne, Direct laser manipulation reveals the mechanics of cell contacts *in vivo*. *Proceedings of the National Academy of Sciences of the United States of America* **112**, 1416–1421 (2015).

58. A. G. Fletcher, F. Cooper, R. E. Baker, Mechanocellular models of epithelial morphogenesis. *Philosophical Transactions of the Royal Society B: Biological Sciences* **372**, 20150519 (2017).

59. M. A. Wyczalkowski, Z. Chen, B. A. Filas, V. D. Varner, L. A. Taber, Computational models for mechanics of morphogenesis. *Birth Defects Research Part C: Embryo Today: Reviews* **96**, 132–152 (2012).

60. G. W. Brodland, How computational models can help unlock biological systems. *Seminars in Cell & Developmental Biology* **47–48**, 62–73 (2015).

61. M. Sadati, A. Nourhani, J. J. Fredberg, N. T. Qazvini, Glass-like dynamics in the cell and in cellular collectives. *WIREs Systems Biology and Medicine* **6**, 137–149 (2014).

62. H. Chaté, F. Ginelli, G. Grégoire, F. Peruani, F. Raynaud, Modeling collective motion: variations on the Vicsek model. *European Physical Journal B* **64**, 451–456 (2008).

63. J. Käfer, T. Hayashi, A. F. M. Marée, R. W. Carthew, F. Graner, Cell adhesion and cortex contractility determine cell patterning in the *Drosophila* retina. *Proceedings of the National Academy of Sciences of the United States of America* **104**, 18549–18554 (2007).

64. S. J. Streichan, M. F. Lefebvre, N. Noll, E. F. Wieschaus, B. I. Shraiman, Global morphogenetic flow is accurately predicted by the spatial distribution of myosin motors. *eLife* **7**, e27454 (2018).

65. B. He, K. Doubrovinski, O. Polyakov, E. Wieschaus, Apical constriction drives tissue-scale hydrodynamic flow to mediate cell elongation. *Nature* **508**, 392–396 (2014).

66. S. Alt, P. Ganguly, G. Salbreux, Vertex models: from cell mechanics to tissue morphogenesis. *Philosophical Transactions of the Royal Society B: Biological Sciences* **372**, 20150520 (2017).

67. D. Bi, J. H. Lopez, J. M. Schwarz, M. L. Manning, A density-independent rigidity transition in biological tissues. *Nature Physics* **11**, 1074–1079 (2015).

68. J. Rozman, M. Krajnc, P. Ziherl, Collective cell mechanics of epithelial shells with organoid-like morphologies. *Nature Communications* **11**, 3805 (2020).

69. D. Bi, J. H. Lopez, J. M. Schwarz, M. Lisa Manning, Energy barriers and cell migration in densely packed tissues. *Soft Matter* **10**, 1885–1890 (2014).

70. D. Bi, X. Yang, M. C. Marchetti, M. L. Manning, Motility-driven glass and jamming transitions in biological tissues. *Physical Review X* **6**, 021011 (2016).

71. J.-A. Park, *et al.*, Unjamming and cell shape in the asthmatic airway epithelium. *Nature Mater* **14**, 1040–1048 (2015).

72. X. Wang, *et al.*, Anisotropy links cell shapes to tissue flow during convergent extension. *Proceedings of the National Academy of Sciences of the United States of America* **117**, 13541–13551 (2020).

73. W. J. Croft, *Under the microscope: a brief history of microscopy* (World Scientific, 2006).

74. C. Dahmann, Ed., *Drosophila: Methods and Protocols* (Springer New York, 2016).

75. J. W. Lichtman, J.-A. Conchello, Fluorescence microscopy. *Nature Methods* **2**, 910–919 (2005).

76. A. Nwaneshiudu, *et al.*, Introduction to confocal microscopy. *The Journal of investigative dermatology* **132**, 1–5 (2012).

77. A. Ustione, D. W. Piston, A simple introduction to multiphoton microscopy. *Journal of Microscopy* **243**, 221–226 (2011).

78. M. Weber, M. Mickoleit, J. Huisken, "Chapter 11 - Light sheet microscopy" in *Methods in Cell Biology*, Quantitative Imaging in Cell Biology., J. C. Waters, T. Wittman, Eds. (Academic Press, 2014), pp. 193–215.

79. D. L. Farrell, O. Weitz, M. O. Magnasco, J. A. Zallen, SEGGA: a toolset for rapid automated analysis of epithelial cell polarity and dynamics. *Development* **144**, 1725–1734 (2017).

80. B. Aigouy, D. Umetsu, S. Eaton, "Segmentation and Quantitative Analysis of Epithelial Tissues" in *Drosophila: Methods and Protocols*, Methods in Molecular Biology., C. Dahmann, Ed. (Springer, 2016), pp. 227–239.

81. B. Aigouy, *et al.*, Cell flow reorients the axis of planar polarity in the wing epithelium of *Drosophila*. *Cell* **142**, 773–786 (2010).

82. C. Y. B. Leung, R. Fernandez-Gonzalez, Quantitative image analysis of cell behavior and molecular dynamics during tissue morphogenesis. *Methods in Molecular Biology* **1189**, 99–113 (2015).

83. T. Zulueta-Coarasa, R. Fernandez-Gonzalez, Dynamic force patterns promote collective cell movements during embryonic wound repair. *Nature Physics* **14**, 750–758 (2018).

84. B. Aigouy, C. Cortes, S. Liu, B. Prud'Homme, EPySeg: a coding-free solution for automated segmentation of epithelia using deep learning. *Development* **147** (2020).

85. L. A. Davidson, Epithelial machines that shape the embryo. *Trends in Cell Biology* **22**, 82–87 (2012).

86. R. Keller, *et al.*, Mechanisms of convergence and extension by cell intercalation. *Philosophical Transactions of the Royal Society of London. Series B: Biological Sciences* **355**, 897–922 (2000).

87. M. Tada, C.-P. Heisenberg, Convergent extension: using collective cell migration and cell intercalation to shape embryos. *Development* **139**, 3897–3904 (2012).

88. A. Mazumdar, M. Mazumdar, How one becomes many: Blastoderm cellularization in *Drosophila melanogaster*. *BioEssays* **24**, 1012–1022 (2002).

89. K. E. Kasza, J. A. Zallen, Dynamics and regulation of contractile actin–myosin networks in morphogenesis. *Current Opinion in Cell Biology* **23**, 30–38 (2011).

90. G. B. Blanchard, *et al.*, Tissue tectonics: morphogenetic strain rates, cell shape change and intercalation. *Nature Methods* **6**, 458–464 (2009).

91. T. Lecuit, P.-F. Lenne, Cell surface mechanics and the control of cell shape, tissue patterns and morphogenesis. *Nature Reviews Molecular Cell Biology* **8**, 633–644 (2007).

92. F. van Roy, G. Berx, The cell-cell adhesion molecule E-cadherin. *Cell. Mol. Life Sci.* **65**, 3756–3788 (2008).

93. G. Jürgens, E. Wieschaus, C. Nüsslein-Volhard, H. Kluding, Mutations affecting the pattern of the larval cuticle in *Drosophila melanogaster*. *Wilhelm Roux's Archives of Developmental Biology* **193**, 267–282 (1984).

94. U. Tepass, *et al.*, *Shotgun* encodes *Drosophila* E-cadherin and is preferentially required during cell rearrangement in the neurectoderm and other morphogenetically active epithelia. *Genes & Development* **10**, 672–685 (1996).

95. T. Lecuit, A. S. Yap, E-cadherin junctions as active mechanical integrators in tissue dynamics. *Nature Cell Biology* **17**, 533–539 (2015).

96. X. Liang, G. A. Gomez, A. S. Yap, Current perspectives on cadherin-cytoskeleton interactions and dynamics. *Cell Health & Cytoskeleton* **7**, 11–24 (2015).

97. T. H. Morgan, The theory of the gene. *The American Naturalist* **51**, 513–544 (1917).

98. Six Nobel prizes – what's the fascination with the fruit fly? *The Guardian* (2017).

99. E. C. R. Reeve, *Encyclopedia of Genetics* (Routledge, 2014).

100. L. T. Reiter, L. Potocki, S. Chien, M. Gribskov, E. Bier, A systematic analysis of human disease-associated gene sequences in *Drosophila melanogaster*. *Genome Research* **11**, 1114–1125 (2001).

101. T. E. Lloyd, J. P. Taylor, Flightless Flies: *Drosophila* models of neuromuscular disease. *Annals of the New York Academy of Sciences* **1184**, e1-20 (2010).

102. U. B. Pandey, C. D. Nichols, Human disease models in *Drosophila melanogaster* and the role of the fly in therapeutic drug discovery. *Pharmacological Reviews* **63**, 411–436 (2011).

103. J. T. Blankenship, S. T. Backovic, J. S. P. Sanny, O. Weitz, J. A. Zallen, Multicellular rosette formation links planar cell polarity to tissue morphogenesis. *Developmental Cell* **11**, 459–470 (2006).

104. K. E. Kasza, D. L. Farrell, J. A. Zallen, Spatiotemporal control of epithelial remodeling by regulated myosin phosphorylation. *Proceedings of the National Academy of Sciences of the United States of America* **111**, 11732–11737 (2014).

105. A. 105ra, *et al.*, A fluid-to-solid jamming transition underlies vertebrate body axis elongation. *Nature* **561**, 401–405 (2018).

106. M. Williams, W. Yen, X. Lu, A. Sutherland, Distinct apical and basolateral mechanisms drive planar cell polarity-dependent convergent extension of the mouse neural plate. *Developmental Cell* **29**, 34–46 (2014).

107. Z. Sun, *et al.*, Basolateral protrusion and apical contraction cooperatively drive *Drosophila* germ-band extension. *Nature Cell Biology* **19**, 375–383 (2017).

108. M. Zhu, *et al.*, Spatial mapping of tissue properties *in vivo* reveals a 3D stiffness gradient in the mouse limb bud. *Proceedings of the National Academy of Sciences of the United States of America* **117**, 4781–4791 (2020).

109. J.-F. Rupprecht, *et al.*, Geometric constraints alter cell arrangements within curved epithelial tissues. *Molecular Biology of the Cell* **28**, 3582–3594 (2017).

110. J. P. Stevens, H. Watanabe, J. W. Kysar, A. K. Lalwani, Serrated needle design facilitates precise round window membrane perforation. *Journal of Biomedical Materials Research Part A* **104**, 1633–1637 (2016).

111. A. Aksit, *et al.*, *In-vitro* perforation of the round window membrane via direct 3-D printed microneedles. *Biomedical Microdevices* **20**, 47 (2018).

112. R. Keller, Physical Biology Returns to Morphogenesis. *Science* **338**, 201–203 (2012).

113. D. Gilmour, M. Rembold, M. Leptin, From morphogen to morphogenesis and back. *Nature* **541**, 311–320 (2017).

114. J. Zhou, H. Y. Kim, L. A. Davidson, Actomyosin stiffens the vertebrate embryo during crucial stages of elongation and neural tube closure. *Development* **136**, 677–688 (2009).

115. L. A. Davidson, Embryo mechanics: balancing force production with elastic resistance during morphogenesis. *Current Topics in Developmental Biology* **95**, 215–241 (2011).

116. O. Campàs, *et al.*, Quantifying cell-generated mechanical forces within living embryonic tissues. *Nature Methods* **11**, 183–189 (2014).

117. G. A. Stooke-Vaughan, O. Campàs, Physical control of tissue morphogenesis across scales. *Current Opinion in Genetics & Development* **51**, 111–119 (2018).

118. A. K. Lawton, *et al.*, Regulated tissue fluidity steers zebrafish body elongation. *Development* **140**, 573–582 (2013).

119. M. Dicko, *et al.*, Geometry can provide long-range mechanical guidance for embryogenesis. *PLOS Computational Biology* **13**, e1005443 (2017).

120. O. Campàs, A toolbox to explore the mechanics of living embryonic tissues. *Seminars in Cell & Developmental Biology* **55**, 119–130 (2016).

121. E. Walck-Shannon, J. Hardin, Cell intercalation from top to bottom. *Nature Reviews Molecular Cell Biology* **15**, 34–48 (2014).

122. A. Vichas, J. A. Zallen, Translating cell polarity into tissue elongation. *Seminars in Cell & Developmental Biology* **22**, 858–864 (2011).

123. M. T. Butler, J. B. Wallingford, Planar cell polarity in development and disease. *Nature Reviews Molecular Cell Biology* **18**, 375–388 (2017).

124. M. Simons, M. Mlodzik, Planar cell polarity signaling: from fly development to human disease. *Annual Review of Genetics* **42**, 517–540 (2008).

125. R. Hale, D. Strutt, Conservation of planar polarity pathway function across the animal kingdom. *Annual Review of Genetics* **49**, 529–551 (2015).

126. C. Bertet, L. Sulak, T. Lecuit, Myosin-dependent junction remodelling controls planar cell intercalation and axis elongation. *Nature* **429**, 667–671 (2004).

127. M. Rauzi, P. Verant, T. Lecuit, P.-F. Lenne, Nature and anisotropy of cortical forces orienting *Drosophila* tissue morphogenesis. *Nature Cell Biology* **10**, 1401–1410 (2008).

128. R. Fernandez-Gonzalez, S. de M. Simoes, J.-C. Röper, S. Eaton, J. A. Zallen, Myosin II dynamics are regulated by tension in intercalating cells. *Developmental Cell* **17**, 736–743 (2009).

129. R. J. Tetley, G. B. Blanchard, A. G. Fletcher, R. J. Adams, B. Sanson, Unipolar distributions of junctional Myosin II identify cell stripe boundaries that drive cell intercalation throughout *Drosophila* axis extension. *eLife* **5**, e12094 (2016).

130. R. Etournay, *et al.*, TissueMiner: A multiscale analysis toolkit to quantify how cellular processes create tissue dynamics. *eLife* **5**, e14334 (2016).

131. L. C. Butler, *et al.*, Cell shape changes indicate a role for extrinsic tensile forces in *Drosophila* germ-band extension. *Nature Cell Biology* **11**, 859–864 (2009).

132. C. M. Lye, *et al.*, Mechanical coupling between endoderm invagination and axis extension in *Drosophila*. *PLOS Biology* **13**, e1002292 (2015).

133. M. Rauzi, *et al.*, Embryo-scale tissue mechanics during *Drosophila* gastrulation movements. *Nature Communications* **6**, 8677 (2015).

134. F. Ag, O. M, B. Re, S. Sy, Vertex models of epithelial morphogenesis. *Biophysical Journal* **106** (2014).

135. T. Nagai, H. Honda, A dynamic cell model for the formation of epithelial tissues. *Philosophical Magazine B* **81**, 699–719 (2001).

136. R. Farhadifar, J.-C. Röper, B. Aigouy, S. Eaton, F. Jülicher, The influence of cell mechanics, cell-cell interactions, and proliferation on epithelial packing. *Current Biology* **17**, 2095–2104 (2007).

137. M. A. Spencer, J. Lopez-Gay, H. Nunley, Y. Bellaïche, D. K. Lubensky, Multicellular actomyosin cables in epithelia under external anisotropic stress. *arXiv:1809.04569 [q-bio]* (2018).

138. D. B. Staple, *et al.*, Mechanics and remodelling of cell packings in epithelia. *European Physical Journal E Soft Matter* **33**, 117–127 (2010).

139. M. Krajnc, S. Dasgupta, P. Ziherl, J. Prost, Fluidization of epithelial sheets by active cell rearrangements. *Physical Review E* **98**, 022409 (2018).

140. L. Yan, D. Bi, Multicellular rosettes drive fluid-solid transition in epithelial tissues. *Physical Review X* **9**, 011029 (2019).

141. D. M. Sussman, M. Merkel, No unjamming transition in a Voronoi model of biological tissue. *Soft Matter* **14**, 3397–3403 (2018).

142. M. Merkel, M. L. Manning, A geometrically controlled rigidity transition in a model for confluent 3D tissues. *New Journal of Physics* **20**, 022002 (2018).

143. M. Merkel, K. Baumgarten, B. P. Tighe, M. L. Manning, A minimal-length approach unifies rigidity in underconstrained materials. *Proceedings of the National Academy of Sciences of the United States of America* **116**, 6560–6568 (2019).

144. A. C. Martin, M. Gelbart, R. Fernandez-Gonzalez, M. Kaschube, E. F. Wieschaus, Integration of contractile forces during tissue invagination. *Journal of Cell Biology* **188**, 735–749 (2010).

145. W. Thielicke, E. Stamhuis, PIVlab – towards user-friendly, affordable and accurate digital particle image velocimetry in MATLAB. *Journal of Open Research Software* **2**, e30 (2014).

146. R. Etournay, *et al.*, Interplay of cell dynamics and epithelial tension during morphogenesis of the *Drosophila* pupal wing. *eLife* **4**, e07090 (2015).

147. M. Merkel, *et al.*, Triangles bridge the scales: Quantifying cellular contributions to tissue deformation. *Physical Review E* **95**, 032401 (2017).

148. M. Leptin, B. Grunewald, Cell shape changes during gastrulation in *Drosophila*. *Development* **110**, 73–84 (1990).

149. C. A. Schneider, W. S. Rasband, K. W. Eliceiri, NIH Image to ImageJ: 25 years of image analysis. *Nature Methods* **9**, 671–675 (2012).

150. D. M. Sussman, cellGPU: Massively parallel simulations of dynamic vertex models. *Computer Physics Communications* **219**, 400–406 (2017).

151. C. Collinet, M. Rauzi, P.-F. Lenne, T. Lecuit, Local and tissue-scale forces drive oriented junction growth during tissue extension. *Nature Cell Biology* **17**, 1247–1258 (2015).

152. J. C. Yu, R. Fernandez-Gonzalez, Local mechanical forces promote polarized junctional assembly and axis elongation in *Drosophila*. *eLife* **5**, e10757 (2016).

153. J. A. Zallen, R. Zallen, Cell-pattern disordering during convergent extension in *Drosophila*. *J. Phys.: Condensed Matter* **16**, S5073–S5080 (2004).

154. M. F. Staddon, K. E. Cavanaugh, E. M. Munro, M. L. Gardel, S. Banerjee, Mechanosensitive junction remodeling promotes robust epithelial morphogenesis. *Biophysical Journal* **117**, 1739–1750 (2019).

155. T. J. C. Harris, Adherens junction assembly and function in the *Drosophila* embryo. *International Review of Cell & Molecular Biology* **293**, 45–83 (2012).

156. Z. Sun, Y. Toyama, Three-dimensional forces beyond actomyosin contraction: lessons from fly epithelial deformation. *Current Opinion in Genetics & Development* **51**, 96–102 (2018).

157. K. E. Cavanaugh, M. F. Staddon, E. Munro, S. Banerjee, M. L. Gardel, RhoA mediates epithelial cell shape changes via mechanosensitive endocytosis. *Developmental Cell* **52**, 152-166.e5 (2020).

158. C. E. Jewett, *et al.*, Planar polarized Rab35 functions as an oscillatory ratchet during cell intercalation in the *Drosophila* epithelium. *Nature Communications* **8**, 476 (2017).

159. A. Sumi, *et al.*, Adherens junction length during tissue contraction is controlled by the mechanosensitive activity of actomyosin and junctional recycling. *Developmental Cell* **47**, 453-463.e3 (2018).

160. G. R. Kale, *et al.*, Distinct contributions of tensile and shear stress on E-cadherin levels during morphogenesis. *Nature Communications* **9**, 5021 (2018).

161. V. D. Varner, J. P. Gleghorn, E. Miller, D. C. Radisky, C. M. Nelson, Mechanically patterning the embryonic airway epithelium. *Proceedings of the National Academy of Sciences of the United States of America* **112**, 9230–9235 (2015).

162. H. Y. Kim, *et al.*, Localized smooth muscle differentiation is essential for epithelial bifurcation during branching morphogenesis of the mammalian lung. *Developmental Cell* **34**, 719–726 (2015).

163. D. M. Alanis, D. R. Chang, H. Akiyama, M. A. Krasnow, J. Chen, Two nested developmental waves demarcate a compartment boundary in the mouse lung. *Nature Communications* **5**, 3923 (2014).

164. Spina bifida. *Centers for Disease Control and Prevention*
www.cdc.gov/ncbddd/birthdefects/surveillancemanual/quick-reference-
handbook/spinaBifida.html (2021).

165. L. Drew, An age-old story of dementia. *Nature* **559**, S2–S3 (2018).

166. F. Peng, *et al.*, Nanoparticles promote *in vivo* breast cancer cell intravasation and extravasation by inducing endothelial leakiness. *Nature Nanotechnology* **14**, 279–286 (2019).

167. A. Jain, R. Bansal, Applications of regenerative medicine in organ transplantation. *Journal of Pharmacy & Bioallied Sciences* **7**, 188–194 (2015).

168. T. E. Yankeelov, *et al.*, Clinically relevant modeling of tumor growth and treatment response. *Science Translational Medicine* **5**, 187ps9-187ps9 (2013).

169. A. R. Westervelt, *et al.*, A parameterized ultrasound-based finite element analysis of the mechanical environment of pregnancy. *Journal of Biomechanical Engineering* **139** (2017).

170. V. Conte, J. J. Muñoz, M. Miodownik, A 3D finite element model of ventral furrow invagination in the *Drosophila melanogaster* embryo. *Journal of the Mechanical Behavior of Biomedical Materials* **1**, 188–198 (2008).

171. What are Rho GTPases? *MBInfo Defining Mechanobiology*.
https://www.mechanobio.info/what-is-mechanosignaling/what-are-small-
gtpases/what-are-rho-gtpases (2021)

172. K. Weigmann, *et al.*, FlyMove – a new way to look at development of *Drosophila*. *Trends in Genetics* **19**, 310-311 (2003).

173. D. Semwogerere, E. R. Weeks, Confocal microscopy. *Encyclopedia of biomaterials and biomedical engineering* **23**, 1-10 (2005).

174. I. Moen, *et al.*, Gene expression in tumor cells and stroma in dsRed 4T1 tumors in eGFP-expressing mice with and without enhanced oxygenation. *BMC Cancer* **12**, 21 (2012).

175. C. Guillot, T. Lecuit, Mechanics of epithelial tissue homeostasis and morphogenesis. *Science* **340**, 1185–1189 (2013).

176. S. Kim, S. Hilgenfeldt, Cell shapes and patterns as quantitative indicators of tissue stress in the plant epidermis. *Soft Matter* **11**, 7270–7275 (2015).

177. G. Grumbling, V. Strelets, The FlyBase Consortium, FlyBase: anatomical data, images and queries. *Nucleic Acids Research* **34**, D484–D488 (2006).

178. J. J. Muñoz, K. Barrett, M. Miodownik, A deformation gradient decomposition method for the analysis of the mechanics of morphogenesis. *Journal of Biomechanics* **40**, 1372–1380 (2007).

179. H. Watanabe, J. W. Kysar, A. K. Lalwani, Microanatomic analysis of the round window membrane by white light interferometry and microcomputed tomography for mechanical amplification. *Otology & Neurotology* **35**, 672–678 (2014).

180. W. B. Shelley, Fluorescent staining of elastic tissue with Rhodamine B and related xanthene dyes. *Histochemie* **20**, 244–249 (1969).

181. S. Wang, T. Hazelrigg, Implications for bcd mRNA localization from spatial distribution of exu protein in *Drosophila* oogenesis. *Nature* **369**, 400–403 (1994).

182. E. Yeh, K. Gustafson, G. L. Boulianne, Green fluorescent protein as a vital marker and reporter of gene expression in *Drosophila*. *Proceedings of the National Academy of Sciences of the United States of America* **92**, 7036–7040 (1995).

183. J. Brasch, O. J. Harrison, B. Honig, L. Shapiro, Thinking outside the cell: how cadherins drive adhesion. *Trends in Cell Biology* **22**, 299–310 (2012).

184. P. Katsamba, *et al.*, Linking molecular affinity and cellular specificity in cadherin-mediated adhesion. *Proceedings of the National Academy of Sciences of the United States of America* **106**, 11594–11599 (2009).

185. J.-L. Maître, *et al.*, Adhesion functions in cell sorting by mechanically coupling the cortices of adhering cells. *Science* **338**, 253–256 (2012).

186. S. Huveneers, *et al.*, Vinculin associates with endothelial VE-cadherin junctions to control force-dependent remodeling. *Journal of Cell Biology* **196**, 641–652 (2012).

187. R. Desai, *et al.*, Monomeric α-catenin links cadherin to the actin cytoskeleton. *Nature Cell Biology* **15**, 261–273 (2013).

188. M. Cavey, M. Rauzi, P.-F. Lenne, T. Lecuit, A two-tiered mechanism for stabilization and immobilization of E-cadherin. *Nature* **453**, 751–756 (2008).

189. N. Borghi, *et al.*, E-cadherin is under constitutive actomyosin-generated tension that is increased at cell–cell contacts upon externally applied stretch. *Proceedings of the National Academy of Sciences of the United States of America* **109**, 12568–12573 (2012).

190. G. A. Gomez, R. W. McLachlan, A. S. Yap, Productive tension: force-sensing and homeostasis of cell-cell junctions. *Trends in Cell Biology* **21**, 499–505 (2011).

191. S. Yonemura, Cadherin-actin interactions at adherens junctions. *Current Opinion in Cell Biology* **23**, 515–522 (2011).

192. N. A. Bulgakova, I. Grigoriev, A. S. Yap, A. Akhmanova, N. H. Brown, Dynamic microtubules produce an asymmetric E-cadherin–Bazooka complex to maintain segment boundaries. *Journal of Cell Biology* **201**, 887–901 (2013).

193. R. Levayer, T. Lecuit, Oscillation and polarity of E-Cadherin asymmetries control actomyosin flow patterns during morphogenesis. *Developmental Cell* **26**, 162–175 (2013).

194. L. A. Perkins, *et al.*, The transgenic RNAi project at Harvard Medical School: resources and validation. *Genetics* **201**, 843–852 (2015).

195. S. van der Walt, *et al.*, scikit-image: image processing in Python. *PeerJ* **2**, e453 (2014).

196. A. A. Hagberg, D. A. Schult, P. J. Swart, Exploring network structure, dynamics, and function using NetworkX in *Proceedings of the 7th Python in Science Conference*, G. Varoquaux, T. Vaught, J. Millman, Eds. (2008), pp. 11–15.

197. H. Ninomiya, *et al.*, Cadherin-dependent differential cell adhesion in *Xenopus* causes cell sorting *in vitro* but not in the embryo. *Journal of Cell Science* **125**, 1877–1883 (2012).

198. M. Tamada, D. L. Farrell, J. A. Zallen, Abl regulates planar polarized junctional dynamics through β-catenin tyrosine phosphorylation. *Developmental Cell* **22**, 309–319 (2012).

199. R. Fernandez-Gonzalez, J. A. Zallen, Oscillatory behaviors and hierarchical assembly of contractile structures in intercalating cells. *Physical Biology* **8**, 045005 (2011).

200. B. J. Bailey, J. T. Johnson, S. D. Newlands, *Head & Neck Surgery--otolaryngology* (Lippincott Williams & Wilkins, 2006).

201. A. Lewis, Cummings otolaryngology: Head and neck surgery, fourth edition. *Head & Neck* **29**, 522–522 (2007).

202. M. A. Garcia, W. J. Nelson, N. Chavez, Cell–cell junctions organize structural and signaling networks. *Cold Spring Harbor Perspectives in Biology* **10**, a029181 (2018).

203. K. Anunts, et al., BigBrain: An ultrahigh-resolution 3D human brain model. *Science* **340**, 1472-1475 (2013).

Appendix A: Supplementary Materials for Chapter 2

A.1 Supplementary materials and methods

A.1.1 Cell shape alignment Q calculation

Cell shape alignment Q was quantified by my collaborators from Lisa Manning's lab using the triangle method following Ref. [147]. A triangular tiling was created based on the barycenters of the cellular polygons, where each vertex of the cellular network gives rise to a triangle whose corners are defined by the barycenters of the three abutting cells. In the case of a manyfold vertex, i.e. a vertex abutting $M > 3$ cells, M triangles are created, where each triangle has one corner defined as the average position of the barycenters of all M abutting cells and the other two corners are the barycenters of two adjacent cells. For each triangle, we computed a symmetric, traceless tensor q quantifying triangle elongation. To compute the tensor q for a given triangle m, we first define a shape tensor s that corresponds to the affine deformation transforming an equilateral reference triangle into the observed triangle m. With the corners of the triangle being at positions r^A, r^B, r^C in counterclockwise order, the shape tensor s can be computed as follows:

$$s = \begin{pmatrix} r_x^B - r_x^A & r_x^C - r_x^A \\ r_y^B - r_y^A & r_y^C - r_y^A \end{pmatrix} \cdot \begin{pmatrix} 1 & 1/2 \\ 0 & \sqrt{3}/2 \end{pmatrix}^{-1}. \tag{A1}$$

From the triangle shape tensor, the triangle elongation tensor q is extracted, which characterizes the anisotropic component of the deformation characterized by s. It is computed by first splitting s into trace part t, symmetric, traceless part $\tilde{s}$, and antisymmetric part s^a:

$$s = t + \tilde{s} + s^a. \tag{A2}$$

Then first a triangle rotation angle θ is extracted such that:

$$\begin{pmatrix} cos\theta \\ sin\theta \end{pmatrix} = a \begin{pmatrix} t_{xx} \\ s_{yx}^a \end{pmatrix} \tag{A3}$$

with some prefactor a. In practice, θ can be extracted using the "arctan2" function that exists in many programming languages as $\theta = \arctan2(s^a_{yx}, t_{xx})$. Finally, the triangle elongation tensor q is computed as:

$$q = \frac{1}{|\tilde{s}|}\operatorname{arcsinh}\left(\frac{|\tilde{s}|}{(\det s)^{\frac{1}{2}}}\right)\tilde{s}\cdot R(-\theta), \tag{A4}$$

where $|\tilde{s}| = \left[s^2_{xx} + s^2_{xy}\right]^{1/2}$ is the magnitude of the symmetric, traceless tensor $\tilde{s}$, $\det s$ is the determinant of the shape tensor s, and $R(-\theta)$ is a clockwise rotation by angle θ:

$$R(-\theta) = \begin{pmatrix} \cos\theta & \sin\theta \\ -\sin\theta & \cos\theta \end{pmatrix}. \tag{A5}$$

The cell shape alignment tensor

$$Q = \begin{pmatrix} Q_{xx} & Q_{xy} \\ Q_{xy} & -Q_{xx} \end{pmatrix} \tag{A6}$$

is then the average of the symmetric, traceless elongation tensors q of all triangles:

$$Q = \langle q \rangle. \tag{A7}$$

The average is an area-weighted average $\langle q \rangle := (\sum_m a_m q_m)/(\sum_m a_m)$, where the sums are over all triangles m with area a_m and elongation tensor q_m. The cell shape alignment parameter Q in the main text of this book is the magnitude of this tensor defined by $Q = \left[Q^2_{xx} + Q^2_{xy}\right]^{1/2}$.

Our cell shape alignment parameter Q combines information about both cell shape anisotropy and cell shape alignment. It can be split accordingly into a product:

$$Q = |\langle q \rangle| = Q_s Q_a. \tag{A8}$$

The first factor $Q_s = \langle |q| \rangle$ is the average magnitude of triangle anisotropy, which is a proxy for cell shape anisotropy, and the second factor $Q_s = \left|\langle \frac{q}{|q|} \frac{|q|}{|\langle q \rangle|} \rangle\right|$ is the norm of the average triangle elongation axis $q/|q|$ weighted by the norm of q. This second factor thus corresponds to an alignment separate from cell shape, which similarly to a nematic order parameter varies between

zero (random shape orientation) and one (perfectly aligned shapes).

A.1.2 Detailed vertex model simulation

The vertex model and the related computational analyses were finished by my collaborators

from Lisa Manning's lab.

Our vertex model describes an epithelial tissue as a planar tiling of N cellular polygons,

where the degrees of freedom are the vertex positions $r_{k\alpha}$ [136]. We use Latin indices starting with

k to refer to vertices and Greek indices starting with α to refer to spatial dimensions. Forces are

defined such that cell perimeters and areas act as effective springs with a preferred perimeter p_0

and a preferred area of one. This is implemented via the following effective energy functional,

which in dimensionless form is [143]:

$$E = \sum_{i=1}^{N}[(p_i - p_0)^2 + k_A(a_i - 1)^2] \tag{A9}$$

Here, the sum is over all cells i, with perimeter p_i and area a_i. The parameter k_A is a

dimensionless number comparing area and perimeter rigidity. We use periodic boundary

conditions with box size $L_x \times L_y$ such that the average cell number density is one: $L_x L_y = N$. The

boundary conditions can accommodate a skew (as in Lees-Edwards boundary conditions) with a

corresponding simple shear γ. Hence, the system energy is a function of all vertex positions and

the periodic box parameters: $E = E(\{r_{k\alpha}\}, L_x, L_y, \gamma)$. We focus on stable, force-balanced states of

the system, which corresponds to local minima of E. To numerically find such states, we use the

BFGS2 multidimensional minimization routine of the Gnu scientific library (GPL) with a cutoff

on the average residual force of 10^{-6}. We allow for manyfold vertices – i.e. vertices are allowed

to be in contact with more than three cells at once. During the minimization, the vertices belonging to an edge are fused to a single vertex whenever the edge length is below a cutoff of 10^{-3}, and a vertex with at least four edges attached to it splits into several vertices whenever this is energetically favorable. While it is known that the existence of manyfold vertices can change the transition point in vertex models [140], we checked that the energy-minimized states we obtained rarely contained any manyfold vertices. In all simulations, we have $N = 512$ cells and $k_A = 1$.

For a given local energy minimum, we compute the simple shear modulus G as described in [142]:

$$G = \frac{1}{N}\left(\frac{\partial^2 E}{\partial\gamma^2} - \sum_m \frac{1}{\omega_m^2}\left[\sum_{k,\alpha}\frac{\partial^2 E}{\partial\gamma\partial r_{k\alpha}}u_{k\alpha}^m\right]^2\right) \tag{A10}$$

In the second term, the outer sum is over all positive eigenvalues ω_m^2 and the corresponding eigenvectors $u_{k\alpha}^m$ of the Hessian matrix $(\partial^2 E/\partial r_{k\alpha}\partial r_{l\beta})$. In practice, we include all eigenvalues smaller than 10^{-14} in the sum. The inner sum in the second term is over all vertices and both spatial dimensions.

A.1.3 Detailed anisotropic vertex model

In all our simulations, we initialize the system with the Voronoi tessellation of a uniformly random point pattern on a squared domain ($L_x = L_y = L_0$). For the first set of simulations of anisotropic tissue (Fig. 2.8a), we apply an external pure shear strain ε by setting $L_x = e^\varepsilon L_0$ and $L_y = e^{-\varepsilon}L_0$. We start with $\varepsilon = 0$ and increase in steps of 0.02 up to a value of $\varepsilon = 2$, minimizing the energy after each step. For these minimizations, we vary all vertex positions, keep the box dimensions fixed, but also allow the simple shear variable γ to vary (shear-stabilized minimization). We follow this protocol for different values of p_0, which we varied between 3.5 and 4.5 in steps of 0.01. For each value of p_0 we carry out 100 separate simulation runs.

For a second set of simulations of an anisotropic tissue (Fig. 2.8b, Inset), we model the anisotropic myosin distribution in the germband by introducing an additional anisotropic line tension with amplitude λ_0 into the effective energy functional:

$$E = \sum_{i=1}^{N}[(p_i - p_0)^2 + k_A(a_i - 1)^2] + \sum_{\langle k,l \rangle} \lambda_{\langle k,l \rangle} \ell_{\langle k,l \rangle} \tag{A11}$$

While the first sum is the same as in Eq. (A9), we have added a second sum, which is over all edges in the system, connecting two vertices k, l. Here, $\ell_{\langle k,l \rangle}$ denotes the length the edge $\langle k, l \rangle$, and $\lambda_{\langle k,l \rangle}$ is a line tension associated with this edge. Before each minimization, we define each of these line tensions based on the respective edge angle $\theta_{\langle k,l \rangle}$ as follows:

$$\lambda_{\langle k,l \rangle} = \lambda_0 \cos(2[\theta_{\langle k,l \rangle} - \phi]) \tag{A12}$$

Thus, the line tension will be increased by λ_0 for edges parallel to lines with angle ϕ and decreased by λ_0 for edges perpendicular to that.

During each minimization run, we vary all vertex positions and the pure shear strain $\varepsilon = 1/2\log(L_x/L_y)$, but keep the system area $L_x L_y$ and the simple shear strain γ fixed. While we set $\lambda_{\langle k,l \rangle}$ before a minimization run and keep it constant during the minimization, the angle $\theta_{\langle k,l \rangle}$ usually changes during the minimization as the vertex positions are varied. As a consequence, the state obtained after the minimization will not correspond to an energy minimum anymore once we update the line tensions $\lambda_{\langle k,l \rangle}$ with the new angles $\theta_{\langle k,l \rangle}$. Thus, to identify a force-balanced state where the line tensions are consistent with the directions of the cell edges, we iterate over several minimizations, where after each minimization we update $\lambda_{\langle k,l \rangle}$ based on the latest angles $\theta_{\langle k,l \rangle}$. We stop these iterations once the states do not significantly change anymore, or more precisely, when the average residual stress per degree of freedom before a minimization, but with the new $\lambda_{\langle k,l \rangle}$, is smaller than 2×10^{-6}. We intentionally do not include the explicit dependency of $\lambda_{\langle k,l \rangle}$ on $\theta_{\langle k,l \rangle}$ and thus the vertex positions in our energy minimizations, because this would create additional torques in our model, while here we merely want to study the effect of an anisotropic distribution of line tensions as provided for instance by an anisotropic myosin distribution.

We set the direction of line tension anisotropy parallel to the y axis, i.e. $\phi = \pi/2$. For Fig. 2.8b, we varied the magnitude of line tension anisotropy λ_0 between zero and one in steps of 0.1, and for Fig. 2.8b inset, we varied it in steps of 0.01. Again, the preferred perimeter p_0 is varied between 3.5 and 4.5 in steps of 0.01, where for each value of p_0 we run 100 separate simulations.

We found many states where the system flowed during a minimization until ε was so large that the system was only one cell thick in the y direction. In particular, this was the case in what was otherwise expected to be the floppy regime. This will probably not only occur in the floppy regime, but also in the solid regime whenever the anisotropic stress created by the line tension anisotropy is large enough to overcome the yield stress, which perhaps explains why there is a gap between mechanically stable solid states and the black line in Fig. 2.8b inset. Because we did not obtain any force-balanced state of bulk vertex model tissue in this regime, we have no way to determine from our simulations neither the shear modulus, nor the morphological quantities $\bar{p}$ and Q, and thus this regime does not appear in Fig. 2.8b inset. To access this regime, one needs to include dynamics into the model, e.g. including a viscosity or a substrate friction.

In order to prevent the system from flowing indefinitely, we also ran a third set of simulations (Fig. 2.8b), where we combined anisotropic line tensions with a fixed system size. In other words, we ran simulations like the second set where we now fixed also the pure shear strain ε, which we successively increased and each time looked for a force-balanced state as described for the second simulation set. While our findings are consistent with our results from the other two simulation sets, we again do not obtain any fluid states for any nonzero value of the line tension anisotropy λ_0, because the iterative procedure involving updating line tensions and minimizing the energy described above does not converge in this regime. To some extent this can even be understood analytically. We start from a fix point $(r^*, \theta^*) = (\{\vec{r}_k^*\}, \{\theta_{(k,l)}^*\})$ of the iterative procedure described above, where $\partial E/\partial r(r^*, \theta^*) = 0$ and $\theta^* = \theta(r^*)$. A small deviation δr_1 from this state in the vertex positions leads to a change in the bond angles of $\delta\theta_1 = \mathrm{d}\theta(r^*)/\mathrm{d}r \cdot \delta r_1$. This leads in turn to an energy minimized state at vertex positions $r_2 = r^* + \delta r_2$ with

$$0 = \frac{\partial E(r^* + \delta r_2, \theta^* + \delta\theta_1)}{\partial r} = \frac{\partial^2 E(r^*, \theta^*)}{\partial r^2} \cdot \delta r_2 + \frac{\partial^2 E(r^*, \theta^*)}{\partial r \partial \theta} \cdot \frac{\mathrm{d}\theta(r^*)}{\mathrm{d}r} \cdot \delta r_1. \tag{A13}$$

Here, $\partial^2 E(r^*, \theta^*)/\partial r^2$ is the Hessian of the system and $\partial^2 E(r^*, \theta^*)/\partial r \partial \theta$ scales linearly with the line tension anisotropy λ_0. According to Eq. (A13), if eigenvalues of the Hessian are sufficiently small that have a nonzero overlap with the second term in Eq. (A13), then any initially small deviation δr_1 from the fix point (r^*, θ^*) will grow during the iterative procedure. In particular, this would explain why we did not observe any stable fluid states for $\lambda_0 > 0$, where the Hessian is expected to contain non-trivial zero modes. We note that this is likely more than just a technical phenomenon related to our quasi-static simulations. This result could instead indicate that generally fluid tissues with anisotropic line tension may not be able to easily attain a stationary state even when boundary conditions prevent overall anisotropic tissue flow.

To obtain Fig. 2.8a, b and inset we binned all of our energy-minimized configurations with respect to $\bar{p}$ and Q (which were computed as described in Section 2.2.1.5 and Appendix A.1.1), and then computed the fraction of floppy configurations within each bin. A configuration was defined floppy when its shear modulus G was below a cutoff value of 10^{-5}.

A.1.4 Analysis of the packing dependence of transition point

To study the packing-dependence of the transition point, we annealed the isotropic vertex model tissue at different temperatures prior to quenching the system to zero temperature, as this is a standard method for altering packing disorder in other materials such as structural glasses. To simulate the vertex model at a given temperature, we followed an Euler integration scheme updating all vertex positions $r_{k\alpha}$ as follows in each time step Δt:

$$r_{k\alpha} \rightarrow r_{k\alpha} + \mu F_{k\alpha}\Delta t + \eta_{k\alpha} \tag{A14}$$

Here, we have non-dimensionalized time such that the dimensionless motility μ is one, $F_{k\alpha} = -\partial E/\partial r_{k\alpha}$ is the force on vertex k with the energy given by Eq. (A9), and $\eta_{k\alpha}$ is a normal distributed random force with zero average and variance $\langle \eta_{k\alpha}\eta_{l\beta} \rangle = 2\mu T\Delta t\delta_{kl}\delta_{\alpha\beta}$. To simulate these dynamics, we use the publicly available cellGPU code [150], with a time step of $\Delta t = 0.01$. For these simulations, vertices are always 3-fold coordinated and an edge undergoes a full T1 transition whenever its length is below a cutoff of 0.04.

We run 100 simulations for each set of parameters (T, p_0), where T varies logarithmically between 5×10^{-6} and 1.5×10^{-1}, and p_0 varies between 3.7 and 3.9 in steps of 0.01. All simulations are thermalized at their target temperature for a time of 10^4 before recording the data. We then perform simulations for 10^6 at the target temperature before the temperature is quenched to $T = 0$. We calculate the decay of the self-overlap function for the slowest of the most solid states (low p_0 and T sets) and confirm that the vertices are displaced less than a characteristic distance of 1/e at time 10^6, suggesting the states are relaxed. To quench the temperature to zero, we run Eq. (A14) for an additional time of 10^6. Afterwards, we use the BFGS2 algorithm of the GSL to further minimize until the average residual force per degree of freedom is below 10^{-6}. The data are shown in Appendix A.2, Fig. A2, demonstrating that the transition point p_0^* does depend systematically on the annealing temperature and therefore on the packing disorder.

The transition point we find occasionally decreases below the value of 3.81 (Appendix A, Fig. A3), which is the minimal transition point we would expect for disordered packings [67]. To test whether this could be due to partial crystallization, we also quantified a hexatic order parameter:

$$\Phi_6 = \frac{1}{N_e}\sum_{\langle k,l \rangle} e^{6i\theta_{\langle k,l \rangle}} \tag{A15}$$

Here, the sum is over all N_e edges in the system, where $\theta_{\langle k,l \rangle}$ is the angle of the edge between vertices k and l. We find that the decreased transition point is indeed correlated with hexatic order $|\Phi_6|^2$ (Appendix A, Fig. A3).

A.1.5 Detailed theoretical expectation for the shift of the transition point

In a recent publication [143], my collaborators' lab showed that the transition point $\overline{p}_{crit}$ in the vertex model is expected to shift away from the isotropic transition point p_0^* as the material is anisotropically deformed with strain $\varepsilon = 1/2\log(L_x/L_y)$ as:

$$\overline{p}_{crit} = p_0^* + 4b\varepsilon^2 \tag{A16}$$

Here, b is a constant prefactor whose precise value depends on the packing disorder, but whose typical value was previously found to be 0.6 ± 0.2 (average $\pm$ standard deviation). We use here the pure shear strain variable $\varepsilon = 1/2\log(L_x/L_y)$, which is related to the strain variable γ used in Ref. [143] as $\varepsilon = \gamma/2$, and so we get an additional factor of 4 in front of b in Eq. (A16). However, Eq. (A16) has so far only been discussed without cell rearrangements, which do occur in our simulations.

To apply these ideas here, we start from an anisotropic configuration that results from some externally applied area-preserving anisotropic strain ε_{xx} that for simplicity we define here to extend the tissue along the x axis. It is possible that cell rearrangements occur during this deformation process, which are not taken into account in Eq. (A16). Thus, we need to disentangle overall strain from the cell rearrangements that it may cause. In Ref. [147], my collaborators have shown before that in the limit of homogeneous deformation without global rotations, the overall strain of a dense 2D cellular network can be decomposed into:

$$\varepsilon_{xx} = Q_{xx}^{\text{final}} - Q_{xx}^{\text{initial}} + \sum_i \Delta X_{xx}^i \tag{A17}$$

Here, Q_{xx}^{initial} and Q_{xx}^{final} are measures for nematic cell shape alignment before and after the deformation process, defined as described in Section 2.2.1.5 and Appendix A.1.1 (Eq. (A1) and following), projected on the x axis. The sum is over all T1 transitions i that occur during the deformation process, where each T1 transition contributes an amount ΔX_{xx}^i to the overall strain. Ref. [147] more generally derives a relation about symmetric, traceless tensors, which we projected onto the x axis for simplicity here.

To apply Eq. (A16) to any anisotropic configuration with nematic cell shape alignment Q_{xx}, we ask for the amount of strain needed to deform this anisotropic configuration into an isotropic one without any cell rearrangements. Defining an isotropic configuration as one where the nematic cell shape alignment is zero $Q_{xx}^{\text{final}} = 0$, we obtain from Eq. (A17) that the strain needed to deform our starting configuration into an isotropic one without T1 transitions is $\varepsilon_{xx} = -Q_{xx}$. Conversely a strain of $\overline{\varepsilon}_{xx} = Q_{xx}$ is needed to get from that isotropic state back to the initial anisotropic state. With p_0^* being the transition point of the isotropic state, we thus obtain from Eq. (A16) that the transition point of our anisotropic tissue state is:

$$\overline{p}_{crit} = p_0^* + 4b\overline{\varepsilon}_{xx}^2 = p_0^* + 4bQ_{xx}^2 \tag{A18}$$

Here we have assumed that all deformations are along the x axis, but the same line of argument applies to any arbitrary axis, such that we finally obtain

$$\overline{p}_{crit} = p_0^* + 4bQ^2 \tag{A19}$$

Here, Q being the magnitude of the nematic cell shape alignment defined in Section 2.2.1.5 and Appendix A.1.1.

An alternative way to obtain Eq. (A19) is to Taylor expand $\overline{p}_{crit}$ in terms of the cell shape alignment tensor $\mathbf{Q}$, where the lowest-order term besides the constant allowed by symmetry is a term $\sim Q^2$. However, with the approach above, we can also connect the value of the prefactor b to previous results.

Finally, the predictions in Ref. [143] strictly speaking refer to the non-dimensionalized average perimeter, i.e. the average of p_i over all cells, whereas here by $\overline{p}$ we refer to the average shape index, i.e. the average of $p_i/\sqrt{a_i}$ over all cells. We verified that this difference does not play a role in our vertex model simulations.

A.1.5.1 Fit to simulation data

To fit Eq. (A19) to the simulation data where we apply the external anisotropic deformation (Fig. 2.8a), we compute the average transition point for each Q by interpreting the $\overline{p}$-dependent fraction of floppy configurations for fixed Q as a cumulative probability distribution function and extracting the average from it. For varying Q, the resulting plot together with a fit to Eq. (A19) is shown in Appendix A.2, Fig. A3.

We excluded a few data points from the fit, which were affected by the excess of rigid states observed around $\overline{p} \approx 4.15$ and $Q \approx 0.3$. This excess of rigid states very likely comes from the occurrence of higher coordinates vertices (Appendix A.2 Fig. A3), which are known to increase the vertex model transition point [140].

A.1.5.2 Fit to experimental data and quality of fit

To compare our theoretical predictions to experimental data, we define a quality of fit measure n_{tot}, which we define as the number of experimental data points that are wrongly categorized as either solid or fluid by our theory. Experimentally, a data point is declared fluid if the instantaneous cell rearrangement rate averaged over a 1.5-minute time interval surpasses a cutoff value, which we set to 0.02 rearrangements per cell and minute for the plots in Chapter 2 Figs. 2.5c, 2.9c, 2.10a-c, 2.12a-c. A few of our theoretical predictions include fit parameters. To determine their value from experimental data, we minimize the quality of fit measure n_{tot} varying those fit parameters.

In Appendix A, Fig. A8, we compare several theoretical predictions by plotting the quality of fit over the rearrangement rate cutoff. To obtain a reliable measurement for the cell rearrangement rate, we fixed the lower limit for the rearrangement rate cutoff requiring that the standard error $2\sigma_r$ of the cell rearrangement rate that we measure is at most as big as its average r. Assuming a Poissonian distribution of the number n_{T1} of cell rearrangements during the measurement interval of $\Delta t = 1.5$ min, we thus find:

$$1 \geq \frac{2\sigma_r}{r} = \frac{2\sigma_{n_{T1}}}{n_{T1}} = \frac{2}{\sqrt{n_{T1}}} = \frac{2}{\sqrt{r\Delta t N_c}} \tag{A20}$$

Hence, to get a reliable measurement for the cell rearrangement rate, we choose a minimal cutoff value of $r_{min} = \frac{4}{\Delta t N_c} = 0.014$ per minute per cell, where N_c is the total number of cells, which is on average $N_c = 190$.

A.2 Supplementary figures

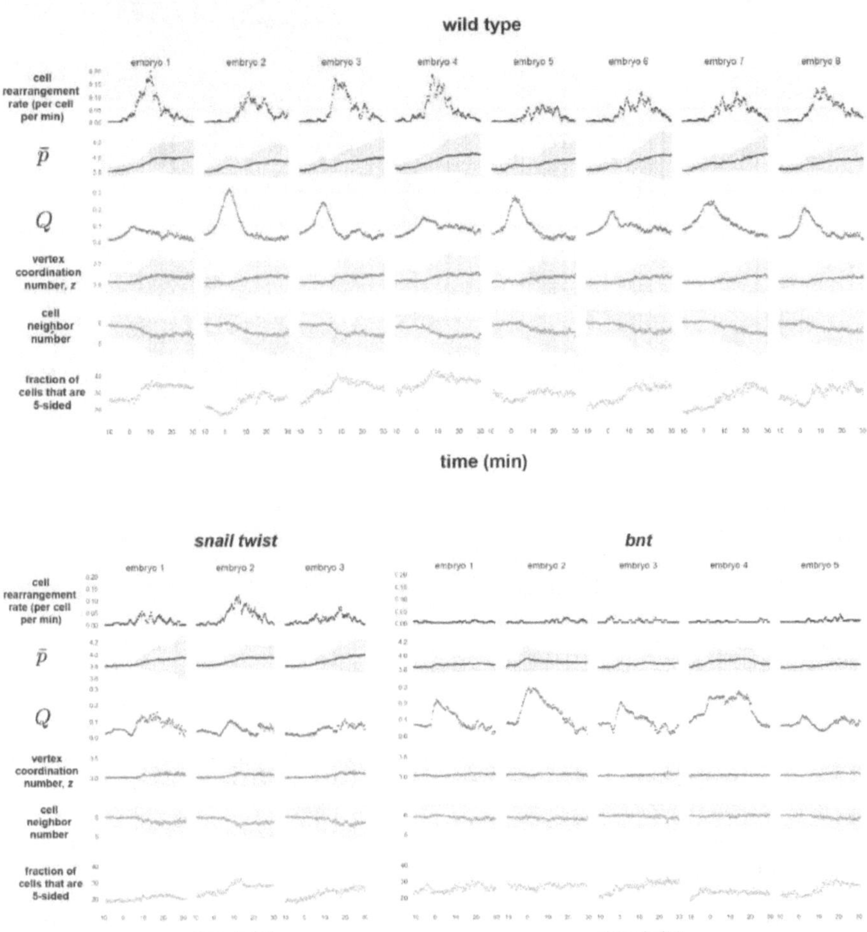

Figure A1. Behavior of the germband in individual wild-type and mutant embryos over time. The cell rearrangement rate per cell per minute, average cell shape index $\bar{p}$, cell shape alignment index Q, average vertex coordination number z, average cell neighbor number, and fraction of cells that are pentagonal f_5.

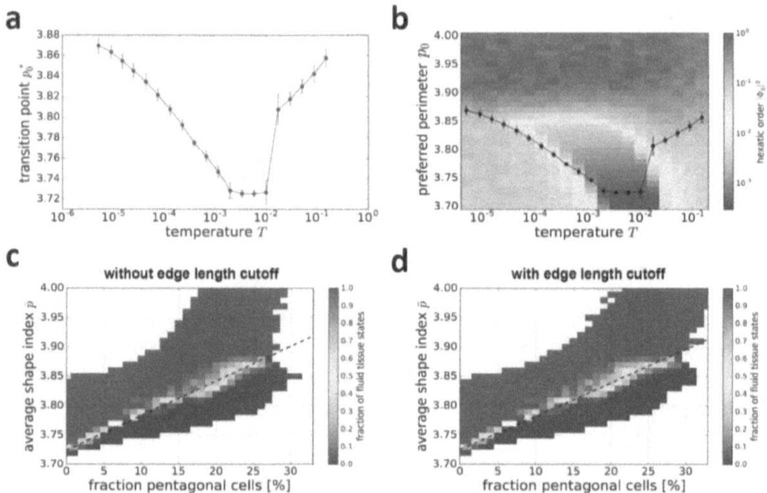

Figure A2. The vertex model transition point depends on the packing disorder. Results of vertex model simulations, where the effect of packing disorder is studied by annealing the model tissue with thermal fluctuations prior to quenching to a force-balanced state to create packings with different degrees of disorder. Taken together, these simulation results show that the critical shape index in the vertex model depends on the cellular packing disorder, where fluctuations can help to decrease packing disorder and p_0^*. (a) The transition point p_0^* decreases with the annealing temperature T before increasing again for very high T, confirming a dependence of the transition point on packing disorder. The values for p_0^* decreased from 3.86 for low temperatures to 3.72 for higher temperatures, and increased again for even higher temperatures. (b) While the lower bound is below the previously reported value of 3.81, this may be related to partial crystallization of the tissue in this regime. The hexatic bond-orientational order parameter, shown here depending on the preferred shape index p_0 and annealing temperature T, indicates at least partial crystallization for intermediate temperatures, which correlates with lower transition points p_0^*. (c) We bin the simulations from panels a, b (100 simulations for each different combination p_0, T) with respect to average cell shape index $\overline{p}$ and fraction of pentagonal cells f_5. Within each bin we then compute the fraction of fluid states using a cutoff of 10^{-7} on the shear modulus. The black dashed line is a linear fit obtained by minimizing the number of simulations on the wrong side of the transition line: $p_0^* = 3.725 + 0.59 f_5$. White regions do not contain any simulation. Same plot as Fig. 2.6a in Chapter 2. (D) Same plot as in panel c with a corrected fraction of pentagonal cells f_5'. As the spatial resolution in our experiments is limited, we detect cell-cell interfaces with a length smaller than $0.11d$ as manyfold vertices, where d is the square root of the average cell area. We applied the same cutoff to interpret the energy-minimized configurations in panel c, which generally leads to a somewhat higher fraction of pentagonal cells $f_5' > f_5$. However, the linear fit of the solid-fluid transition (black dashed line) is only slightly altered to: : $p_0^* = 3.726 + 0.57 f_5'$.

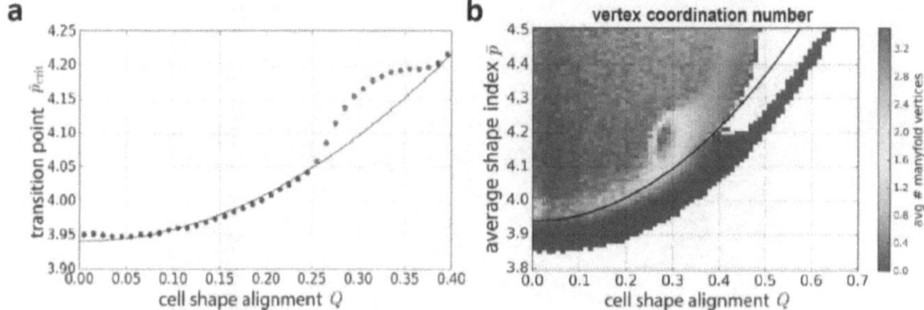

Figure A3. Fits of Eq. 2.2 to simulations. (a) Fit of Eq. 2.2 (black line) to the vertex model simulation results for the case of external deformation (dots, cf. Fig. 2.8a). The average transition points $\overline{p}_{crit}$ were extracted from the simulation data (Fig. 2.8a) by interpreting the fraction of fluid networks for fixed Q as a cumulative probability density and extracting the average from it [142]. From the fit we find $p_0^* = 3.94$ and $b = 0.43$. The red data points were excluded from the fit. These points are related to an excess number of rigid states in the region (cf. panel b and Fig. 2.8a). (b) The excess number of rigid states around $Q \approx 0.3$ and $\overline{p} \approx 4.15$ (cf. panel a and Fig. 2.8a) is associated with an increase in the number of manyfold coordinated vertices in model tissues. Plotted here is the total number of manyfold vertices in configurations of 512 cells.

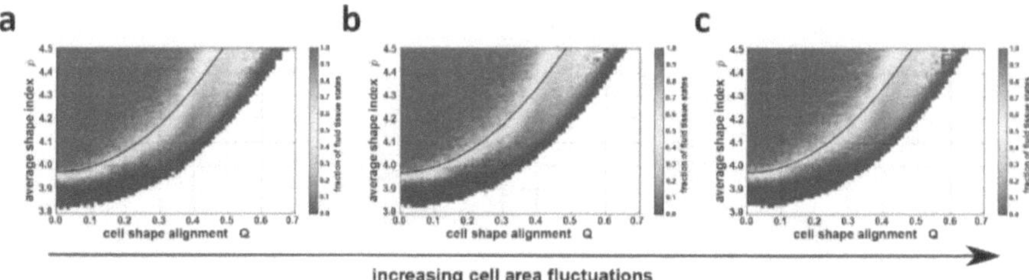

Figure A4. Effects of cell area variation in vertex model simulations. Effects of cell area variation on vertex model simulation results. A Gaussian distribution for the preferred cell areas was used with relative standard deviations of 0% (a), 10% (b), and 20% (c). Solid black lines represent fits to the quadratic relation Eq. 2.2. Variation in cell area has only a very small effect on our theoretical findings and the parameters p_0^* and b. We find the following fit parameters: (a) $p_0^* = 3.96$, $b = 0.57$; (b) $p_0^* = 3.97$, $b = 0.58$; (c) $p_0^* = 3.97$, $b = 0.57$. These parameters differ somewhat from what we find in Figs. 2.8a, A.3a, because there we allowed for manyfold vertices as intermediate states during the energy minimization, whereas in these simulations, we did not allow manyfold vertices in order to reduce simulation run time.

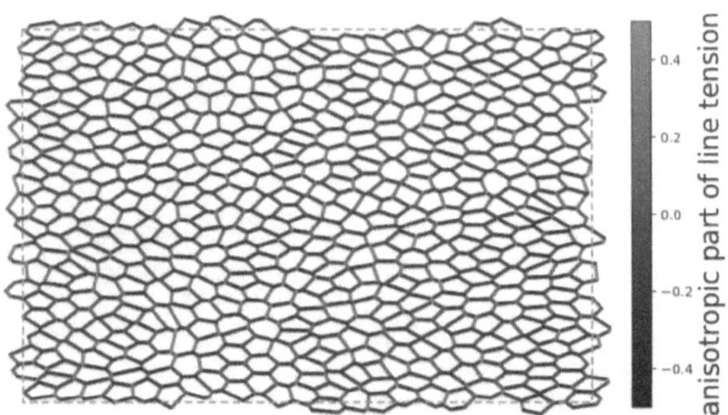

Figure A5. Vertex model tissue with anisotropic internal stresses. Force-balanced vertex model state with anisotropic cell-cell interfacial tensions to model the effects of planar polarized myosin II localization patterns. Cell edges color-coded by the tension anisotropy. Parameters: $p_0 = 3.5, \lambda_0 = 0.5$.

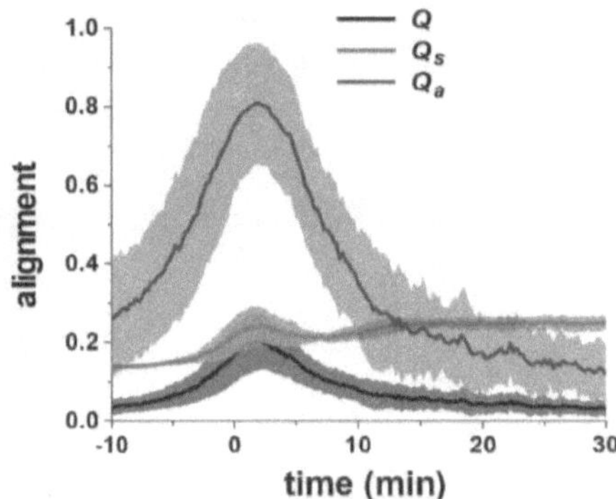

Figure A6. Cell shape alignment. In Chapter 2, we discuss the parameter Q (black curve), which describes both cellular shape anisotropy and alignment of cell shapes in the tissue. Indeed, Q can be decomposed as $Q = Q_s Q_a$. (cf. Eq. (A8)), where Q_s represents cellular shape anisotropy (red curve) and Q_a represents cell shape alignment (blue curve). The "pure" nematic alignment parameter Q_a ranges from zero (random cellular orientation) to one (all cells aligned along the same axis, although with potentially different magnitudes). Here we plot Q, Q_s, and Q_a for the wild-type germband. Q_a is almost one at around $t = 0$, which indicates almost perfect alignment of the cells at that time point.

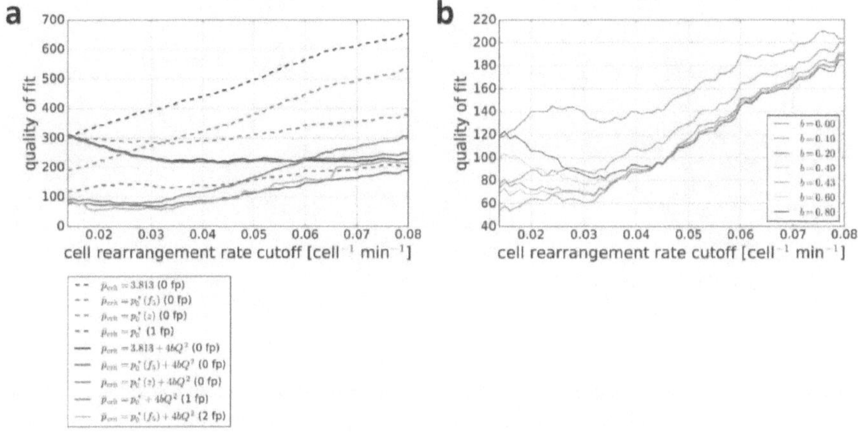

Figure A7. Quality of fits. To compare theoretical predictions to experimental data, we define a quality of fit measure n_{tot}, which is the number of experimental data points that are wrongly categorized as either solid or fluid by the prediction. Thus, a better fit is associated with a smaller value of n_{tot}. (a) Comparison of several theoretical predictions by plotting the quality of fit, n_{tot}, over a range of cell rearrangement rate cutoffs. The predictions include: a constant value of $p_0^* = 3.813$ (black dashed line) [67], a constant value of $p_0^* = 3.813$ and Q-dependent correction according to Eq. 2.2 in Chapter 2 (black solid line), fraction-of-pentagon-dependent $p_0^*(f_5)$ extracted from vertex model simulations in Fig. 2.6a (green dashed line), same with Q-dependent correction (green solid line), vertex-coordination-dependent $p_0^*(z)$ according to Ref. [140] (red dashed line), and same with Q-dependent correction (red solid line; cf. Eq. 2.3 and Fig. 2.10a). Note that all of these six predictions are parameter free ("fp" in the legend indicates the respective number of fit parameters used). In addition, we plot the quality of one-parameter fits, where we either fit a constant p_0^* (blue dashed line; cf. Fig. 2.9c) or additionally include the Q-dependent correction (blue solid line; cf. Fig. 2.9c). Finally, we plot the quality of a two-parameter fit, where for each cell rearrangement cutoff we extract the fraction-of-pentagon dependence $p_0^*(f_5)$ from experimental data using a fit like Fig. 2.10b and include the Q-dependent correction of Eq. 2.2 (cyan solid line; cf. Appendix A.2 Fig. A8). We find that including the Q-dependent correction always improves the prediction, where each time we have fixed $b = 0.43$. Moreover, the best parameter-free prediction for small rearrangement rate cutoffs is given by the red solid line corresponding to Eq. 2.3 in Chapter 2. (b) Quality of fit, n_{tot}, for different fixed values of the parameter b for one-parameter fits of Eq. 2.2 where we vary p_0^* (cf. blue dashed and solid lines in panel a and Fig. 2.9c). Throughout the book, we used the value of $b = 0.43$ determined from vertex model simulations results. The quality of fit is only slightly improved by changing the value of b.

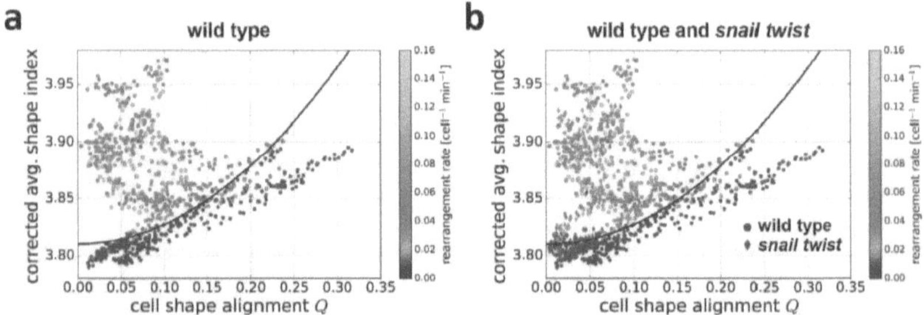

Figure A8. Accounting for packing disorder using the fraction of pentagonal cells. In Chapter 2, we account for the effects of packing disorder on p_0^* by taking into account the vertex coordination number in the tissue (cf. Fig. 2.10c) using the previous prediction in Ref. [140], as shown in Eq. 2.3 in Chapter 2. Alternately, one could use the fraction of pentagonal cells, which in our experiments correlates well with p_0^* (cf. Fig. 2.10b). Here we show the results if we instead correct the cell shape index by the fraction of pentagonal cells at the transition point, according to the linear fit in Fig. 2.10b. (a) Same experimental data as in Fig. 2.10a, but with the cell shape corrected by fraction of pentagonal cells observed in the tissue at the transition point. (b) Same experimental data as in Fig. 2.12b, but with the cell shape corrected by fraction of pentagonal cells observed in the tissue at the transition point.

Appendix B: Supplementary Materials for Chapter 3

B.1 Supplementary protocols

B.1.1 Transparent apple juice plate protocol

This protocol is to make 500 mL of ingredients to make about 50 transparent apple juice plates for *Drosophila* culture, embryo harvest, and experiment conduct, modified from the recipe shared by Nate Hill.

Ingredients:

H_2O	375 mL	Sugar	11.25 g
Apple Juice	125 mL	Tegosept	10 mL
Agar	13.75 g		

1. Tegosept preparation

Dilute tegosept to 10% (e.g. add 200g of the powder to 2000 mL of 190 proof alcohol)

2. Apple juice plates making

(1) Add water 375mL.

(2) Add agar 13.75g with whisk.

(3) Turn on the heat until the agar boils.

* Note: When the agar is done cooking is a personal preference. Sometimes it does not boil. Sometimes it just gets thick. Then it is done when the thickness is good meaning it is not too thick to pour the plates. We can also heat it by the microwave oven for about 90 seconds.

(4) When the agar is done, add 11.25g sugar and let it dissolve in the hot or boiling agar. Whisk it for about 15 seconds till it dissolves.

(5) Turn off the heat.

(6) Add the apple juice 125mL and tegosept 10mL at the same time.

(7) Pour the liquid into petri dishes properly.

(8) Wait for the liquid to solidify and cool down.

B.1.2 *Drosophila* embryo fixation protocol

This protocol is to fix *Drosophila* embryos with formaldehyde and stain E-cadherin and F-actin, modified from the protocol shared by Dr. Jennifer Zallen.

1. Prepare scintillation vials with fixative

Add 100 µL 37% formaldehyde, 900 µL 0.1M sodium phosphate (pH 7.4), and 1mL heptane into the vial to form an effective 3.7% formaldehyde fixative. An interface should form between heptane (top) and sodium phosphate buffer (bottom). If it does not, discard the solution and repeat.

2. Prepare embryos

Dechorionate *Drosophila* embryos for 2 minutes in 50% bleach. Collect embryos in a white wire basket and wash with water for 2 minutes. Blot bottom of basket to dry and transfer embryos with soft paintbrush into the bottom layer of fixative. The embryos will remain at the interface between heptane and sodium phosphate buffer.

3. Formaldehyde fixation

(1) Fix embryos in the 3.7% formaldehyde fixative on a shaker for 1 hour.

(2) Prepare large plates and by covering them with two layers of heptane-glue. To make the glue, fill a scintillation vial with heptane up to a third of the vial and stick double-sided tape inside. Let it sit around for two hours and then suck up the heptane (glue).

(3) Remove all the liquid from the "shaken" embryos, leaving only the interface.

(4) Wash embryos with water into a white mesh (treated as formaldehyde trash). Use only a bit of water at a time.

(5) Keep rinsing the embryos (treated as formaldehyde trash) very thoroughly to get rid of all the heptane.

(6) Rinse the embryos a bit more in the sink.

4. Devitellinize fixed embryos (hand-peeling)

(1) Brush the embryos onto the plates. Spread them well, making sure they stick to the tape. This is key to be able to devitellinize them. If they still have heptane they will not stick.

(2) Cover the plates with PBTriton.

(3) At the microscope, lightly push the embryos from head or tail with a needle. The easiest way to do this is to make an incision with the needle across the embryo's head and then pool the embryo outside the vitelline membrane from behind. The vitelline membrane will then be seen as a transparent layer.

(4) Shake circularly very gently on a black surface to concentrate embryos at the center.

(5) Use Pasteur pipette to move embryos from plate into an eppendorf tube.

(6) Check under the microscope for embryos that are still on the plate and aspirate with glass pipettes.

(7) Remove solution, add blocking solution and continue staining from there.

5. Antibody staining

(1) Block in more than 1 ml block solution for at least 1 on rotating surface at room temperature or overnight on a rotating surface at 4°C.

(2) Incubate with 50 μL primary antibodies diluted in antibody buffer for 2 hours at room temperature. Flick tube occasionally. In Chapter 3, the primary antibody was rat anti-DE-cadherin

DCAD2 (Developmental Studies Hybridoma Bank, DSHB) and used as 1:25 (2 µL DCAD2 antibody and 50 µL antibody solution).

(3) Rinse embryos 2X with> 1 mL PB Triton.

(4) Wash embryos 6X with> 1 mL PB Triton over 2 hours at room temperature or overnight on a rotating surface at 4°C.

(5) Incubate with 300 µL secondary antibody for 45 minutes at room temperature. In Chapter 3, the goat anti-rat secondary antibody conjugated to Alexa-488 was used at 1:500 (Molecular Probes), and Alexa-647-conjugated phalloidin was used at 1:100 (Molecular Probes).

(6) Rinse embryos 2X with> 1 mL PB Tween.

(7) Wash embryos 6X with> 1 mL PB Tween over 2 hours at room temperature or overnight on a rotating surface at 4°C.

(8) Collect embryos from the bottom of the tube and deposit on a glass slide, absorb some of the extra PB Tween solution and mount in 40 µL of ProLong Gold Antifade Mountant. Add a 25 x 40 cover slip and tap very lightly around the embryos to break surface tension for spreading the mountant evenly.

(9) Keep slides at room temperature overnight and then keep at 4°C for long-term storage.

Solutions:

Solutions are made up with MQ water. All solutions containing BSA should be stored at 4°C.

PB Triton (500ml): 1X PBS (478.33ml) + 1% BSA 30% (16.67ml) + 0.1% Triton 10% (5ml)

PB Tween (500ml): 1X PBS (478.33ml) + 1% BSA 30% (16.67ml) + 0.1% Tween 10% (5ml)

Block solution (250ml): 1X PBS (161.7ml) + 10% BSA 30%(83.3ml) + 0.1 % Triton 10% (2.5ml) + 0.1 % azide 10% (2.5ml)

Antibody solution (50ml): 1X PBS (40.67ml) + 5% BSA 30% (8.33ml) + 0.1 % Triton 10% (500µl) + 0.1 % azide 10% (500µl)

Triton 10% stock (10ml): Triton 100 (1ml) + water (9ml) + store in glass

Tween 10% stock (10ml): Tween 20 (1ml) + water (9ml) + store in glas

B.1.3 *Drosophila* DNA sequencing protocol

This protocol is to sequence DNA in *Drosophila* embryos by polymerase chain reaction (PCR). We used this protocol to test if mCherry was in the stock.

1. Fly DNA Preparation for Sequence Protocol

(1) Freeze 3-4 flies (no sex preference) in -20°C for each genotype in a 1.5ml Eppendorf tube. Remember to disinfect the platform and brush with 70% Ethanol and put a Kimwipe under when picking up flies.

(2) Prepare squishing buffer (SB).

       10mM Tris PH8.2

       1mM EDTA

       25mM NaCl

(3) Add final concentration 200ug/ml proteinase K (20mg/ml) freshly from a frozen stock.

(4) Add SB to the proteinase K and make the liquid volume to 100µl in total (99 µl SB + 1µl proteinase K-20mg/ml).

(5) Add 50µl SB-proteinase liquid to each fly Eppendorf tube.

(6) Smash the flies completely with the pipette tip.

(7) Incubate at 37°C for 45min.

(8) Inactivate the proteinase K at 100C for 2min. Fix Eppendorf tube covers by clips in case the covers jump up while being heated.

(9) 13000rpm for 30sec. Remember to keep the centrifuge a balanced weight distribution.

(10) Pipette 30µl of the solution to a new 1.5ml Eppendorf tube, mark it and keep it at -20°C (OK for months).

2. PCR protocol to work with Platinum Pfx Polymerase

(1) Keep all reagents in ice while preparing the reaction mix

	Volume for 50ul reaction	Volume for 10ul reaction
10X Pfx Amplification buffer	5	1
10mM dNTP	1.5	0.3
50mM MgSO4	1	0.2
Primer mix 10µM each	1.5	0.3
Template (10pg ro 200ng)		
Platinum Pfx Polymerase	0.4	0.1
Autoclaved MiliQ Water	to 50 µL	to 10 µL

The volumes provided are for a single reaction. Prepare a master mix of common components for multiple reactions

(2) Gently mix the reaction and quick spin to collect all sample at the bottom.

(3) Transfer PCR tubes from ice to a PCR machine with the block preheated to 98°C and begin thermocycling.

(4) Maintain the reaction at 4°C after cycling. Samples can be stored at −20°C until use.

(5) Analyze the products by agarose gel electrophoresis.

Step		T (°C)	Time
Initial denaturation		94	5min
25-35 cycles	Denature	94	15s
	Anneal	55-72	30s
	Extend	68	1min per Kb
Final extension		68	5min

*Here I used 10μl of volume for reaction. I need 9 reactions, so I made a total of 100μl of master mix and chose the volume I need to each DNA vial. After that I added the template to each DNA vial and started PCR.

3. PCR Protocol for Taq DNA Polymerase with ThermoPol® Buffer (M0267)

This protocol can be found at https://www.neb.com/protocols/0001/01/01/taq-dna-polymerase-with-thermopol-buffer-m0267.

4. Gel electrophoresis protocol

Preparing the gel

(1) Measure 0.3 g of agarose (this is for one minigel, adjust quantities accordingly depending on size or number of gels planned).

(2) Measure 30 mL 1xTAE in a microwavable flask. Add 3ul of SYBR Safe stain 10000X. Mix well and add to the powdered agarose.

(3) Microwave for 1-3 min in 30s pulses-swirl until the agarose is completely dissolved (do not overboil the solution)

(4) Let agarose solution cool down to about 50°C (about 5 min). It can be kept in the water bath at 50C if multiples gels are planned.

(5) Pour the agarose into a gel tray with the well comb in place (slowly to avoid bubbles)

(6) Let the gel solidify at RT (approx. 30 min). If in a hurry put it at 4°C

Loading the gel

(7) Add TrackIT Cyan/Yellow loading buffer to each of your digest samples as follows:

TrackIt™ Cyan/Yellow Loading Buffer (6X)	3.3 µL
DNA Sample	x µL
Deionized Water	to 20 µL

* Here I used 15 wells. So I chose half of the value above. I put the loading buffer in a line on the film and then added 10 µ l of samples to each buffer drop (I changed the tips every time). When loading, I only chose 5 µ l of each to load into the wells.

(8) Once solidified, place the agarose gel into the gel box (electrophoresis unit).

(9) Fill gel box with 1xTAE until the gel is covered.

(10) Load the molecular weight ladder (TrackIT 1Kb Plus DNA ladder 5ul) into the first lane of the gel. (TrackIT comes ready to use)

(11) Carefully load your samples into the additional wells of the gel.

(12) Run the gel at 50-80 V until the dye line is approximately 75-80% of the way down the gel.

* Here I used 65V for 45 min.

* **Note:** Black is negative, red is positive. (The DNA is negatively charged and will run towards the positive electrode.) **Always Run to Red.**

* **Note:** A typical run time is about 1-1.5 hours, depending on the gel concentration and voltage.

(13) Turn OFF power, disconnect the electrodes from the power source, and then carefully remove the gel from the gel box

(14) Using any device that has UV light, visualize your DNA fragments.

(15) For each sample you want to load on a gel, make 10% more volume than needed because several microliters can be lost in pipetting. For example, if you want to load 1.0 μg in 10 μL, make 1.1 μg in 11 μL.

5. The Primers

FORWARDS

SEQUENCE 5'- GGC GAG GAG GAT AAC ATG GCC ATC ATC AAG GAG -3'

LENGTH 33

GC CONTENT 54.5 %

MELT TEMP 65.5 °C

MOLECULAR WEIGHT 10285.7 g/mole

EXTINCTION COEFFICIENT 339800 L/(mole·cm)

nmole/OD260: 2.94

μg/OD260: 30.27

REVERSE

SEQUENCE 5'- GCG TTC GTA CTG TTC CAC GAT GGT GTA GTC CTC -3'

GC CONTENT 54.5 %

MELT TEMP 65.1 °C

MOLECULAR WEIGHT 10102.6 g/mole

EXTINCTION COEFFICIENT 303100 L/(mole·cm)

nmole/OD260: 3.3

µg/OD260: 33.33

Expected Sequence for the mCherry

atggtgagcaagggcgaggaggataacatggccatcatcaaggagttcatgcgcttcaaggtgcacatggagggctccgtgaacggcca

cgagttcgagatcgagggcgagggcgagggccgcccctacgagggcacccagaccgccaagctgaaggtgaccaagggtggccccc

tgcccttcgcctgggacatcctgtcccctcagttcatgtacggctccaaggcctacgtgaagcaccccgccgacatccccgactacttgaag

ctgtccttccccgagggcttcaagtgggagcgcgtgatgaacttcgaggacggcggcgtggtgaccgtgacccaggactcctccctgcag

gacggcgagttcatctacaaggtgaagctgcgcggcaccaacttcccctccgacggccccgtaatgcagaagaagaccatgggctggga

ggcctcctccgagcggatgtaccccgaggacggcgccctgaagggcgagatcaagcagaggctgaagctgaaggacggcggccacta

cgacgctgaggtcaagaccacctacaaggccaagaagcccgtgcagctgcccggcgcctacaacgtcaacatcaagttggacatcacct

cccacaacgaggactacaccatcgtggaacagtacgaacgcgccgagggccgccactccaccggcggcatggacgagctgtacaag

B.1.4 Fly stock quarantine protocol

This protocol is to quarantine fly stocks to avoid potential pests such as mites or molds. Newly requested fly stocks are required to be quarantined.

When getting new fly stocks from outside sources, we need to quarantine the new stocks in case there are mites in them. This is a one-month routine.

1. Check the situation of newly arrived stocks.

2. Flip flies from newly arrived stocks to new B food vials. Write down the date (date 1) and the stock name on each of the new vials. Put them in a new empty plate (plate 1). Put the plate in a separated room from the lab. Check plate 1 for mites one month later.

3. One day after date 1, flip all the flies to new B food vials. Write down the date (date 2) and the stock name on each of the new vials. Put them in a new empty plate (plate 2). Put the plate in a separated room from the lab and away from plate 1.

4. One day after date 2, flip all the flies to new B food vials. Write down the date (date 3) and the stock name on each of the new vials. Throw all the old vials in plate 2. Put new vials plate 2.

5. One day after date 3, flip all the flies to new B food vials. Write down the date (date 4) and the stock name on each of the new vials. Throw all the old vials in plate 2. Put new vials plate 2.

6. If you flip them in public lab space, use 70% ethanol to clean the space in case there are mites.

7. One month after date 1, check if there are mites in vials in plate 1. If there are not, the new stocks in plate 2 are safe to use. If there are, mites cleaning steps are needed.

B.1.5 Pierce BCA protein assay protocol

This protocol is to quantify total proteins. Introductions of the protein assay kit we used can be found at https://www.thermofisher.com/order/catalog/product/23225#/23225.

1. Gather

    Pierce BCA Protein Assay Kit (Thermo Scientific)

    Protein samples

    Diluent: the same buffer used in samples or Milli-Q water

        for *Drosophila* E-Cad level tests, we used 2X LI-COR protein loading buffer

Pipettes and tips

Water bath

Eppendorf tubes

PCR tubes

Nano Drop Machine or Absorbance Microplate Reader

2. Preparation of Standards and Working Reagent (WR)

(1) Make everything **Triplicates** so that we can do averaging in the end.

(2) Standards making: when using NanoDrop, we need 1/10 volumes of standards as in the manual.

Vial #	Volume of Diluent (µl)	Volume and Source of BSA (µl)	Final BSA Concentration (mg/ml)
A	0	30 of stock	2
B	12.5	37.5 of stock	1.5
C	32.5	32.5 of stock	1
D	17.5	17.5 of B	0.75
E	32.5	32.5 of C	0.5
F	32.5	32.5 of E	0.25
G	32.5	32.5 of F	0.125
H	40	10 of G	0.025
I	40	0	0 = Blank

* Volume of diluent: 300 µl.

* If using 4X LI-COR protein loading buffer, dilute 150 µl 4X buffer with 150 µl Milli-Q water.

* When BSA is open, keep it sealed in an Eppendorf tube. It can be used within 3 months.

(3) Determine the total volume of WR required:

(# standards + # unknowns) x (# replicates) x (volume of WR per sample) = WR volume

Example: if we have 4 unknown samples, here we also have 9 standard samples and do triplicates (3 replicates), for each sample 200 µl of WR is needed, then the total volume of WR required is:

(4+9) x 3 x 200 µl = 7800 µl $\approx$ 10000 µl

10000 µl is easier to make. We use 10000 µl of reagent A + 200 µl of reagent B

Reagent A:B = 50:1

3. Sample loading

(1) Turn on the water bath, and set it to 37 °C in advance.

(2) Place PCR tubes in order and mark them with A, B, C, … I, S1, S2, … as below. Record corresponding S1, S2, … samples.

	A	B	C	D	E	F	G	H	I	S1	S2	S3	S4
#1													
#2													
#3													

(3) Each load contains 200 µl of WR + 25 µl of samples.

(4) Load 200 µl of WR to each sample first.

(5) Load 25 µl of standard and unknown samples to each lane, changing the tip among samples.

4. Incubate at 37 °C for 30 min in water bath.

5. Read the samples at Nano Drop Machine to get the protein concentration (mg/ml).

(1) Set up the machine in advance because it needs time to initialize.

(2) Read at 562 nm.

(3) New Experiment -> Proteins -> Protein Pierce 660 -> Curve Type: 2nd order polynomial ->
Replicates: 3

(4) Input standard sample values (the system can record more than 8 groups of data).

(5) When loading, using pipette to drop 2 μl of sample on the platform. **Clean** both the stand and
cover before each load.

(6) Load standard samples (write the data down in notebook because this part of data will not be
able to export).

(7) Load unknown samples, and read the protein concentration.

(8) Hand record or export the unknown sample data using a USB driver.

OR

**5. Read the samples at Absorbance Microplate Reader to get the protein concentration
(mg/ml).**

B.1.6 Western Blots for *Drosophila* embryos protocol

This protocol is to do Western Blots for proteins in *Drosophila* embryos. This protocol is modified
and optimized to do E-Cadherin Western Blots.

1. Materials:

Fly cups

50% bleach

Mesh

Paintbrush

Ice

4X LDS sample buffer w/ BME added just before use (make 300uL of 2X)

      2x solution should be 5% BME (15uL BME + 150ul 4X LDS + 135uL water)

      1x solution should be 2.5% BME (15uL BME + 150ul 4X LDS + 435uL water)

Disposable pestles

Locking eppendorf tubes

NuPAGE Novex Tris-Acetate 3-8% Gel w/ Tris-Acetate SDS running buffer, 10 lanes

      (choose gel depending on protein of interest)

XCell SureLock Mini-Cell

NuPage Transfer buffer 20X, add methanol, antioxidant, water

Nitrocellulose membrane, blot paper, filters

Wet transfer stuff

Ice insert

Bio-Rad Blocking nonfat milk

TBST

Primary antibody solution, TBST + 5% BSA (add sodium azide if storing)

GE Healthcare, Amersham, ECL Plus Western Blotting Detection Reagents RPN2132

Parafilm

HRP secondaries, in glycerol stocks in -80C

Primary antibodies, for example:

- Roche Ms anti-GFP (use at 1:1000) --> ~30uL in MsAb box

- DSHB Ms anti-beta tubulin E7 sup (use at 1:500) --> ~ 200uL in MsAb box

- If we do E-Cadherin Western Blots:

    o DSHB anti-E-Cadherin DCAD2 (use at 1:100)

    o DSHB anti-beta tubulin E7 (use at 1:100) --- control group

2. Day 1: Collect embryos, lyse, freeze

Prepare LDS:

Mix 4X LDS sample buffer w/ BME added just before use (make 300uL of 2X)

  2x solution is 5% BME (15uL BME + 150ul 4X LDS + 135uL Milli-Q water)

Here for LDS sample buffer: use 4X LI-COR Protein Loading Buffer or 4X Laemmli Sample Buffer instead.

For cups at 18C:

(1) Collect for 3 hours at 18°C.

(2) Age plates for 3 hours at 18°C.

(3) Label Eppendorf tubes and put on ice.

(4) Bleach embryos, wash extensively.

(5) Under scope select 150 embryos, use paintbrush to place in eppendorf tube on ice. (This is more than you need for one experiment. You are actually aiming at for 15 embryos loaded per lane.)

(6) Add 75µL 2X LDS buffer.

(7) Homogenize with pestle while on ice.

(8) Spin down 13,000g for 5 min.

(9) Put clamp on tubes and put at 70°C in heat block for 10 min.

(10) Spin 13,000g for 5 min.

(11) Aliquot 25 µl per tube (~30 embryos) for 2 tubes, any remaining fluid in 3rd tube.

(12) Freeze at -20°C for westerns later.

(13) Can add more LDS just before running gels to make sure protein is reduced.

3. Day 2: Run gel and transfer

(1) Turn on heat block to 70°C.

(2) Prepare 1L of 1X Novex transfer buffer and put in cold room.

(3) Prepare 1L of 1X Tris-Acetate SDS running buffer.

(4) Prepare 300uL fresh 1x LDS Sample buffer.

(5) Get XCell Surelock Mini-Cell rinsed and ready.

(6) Mix fresh 1X LDS sample buffer and each sample, aiming to **20 µl (15 embryos)** per well.

(7) Heat at **70°C for 10 min**. Spin down **13k for 5min**.

(8) Prepare the gel, add proper volume of running buffer.

(9) Load **20 µl** sample per well.

(10) Load **5 µl** of the LI-COR Chameleon Duo Pre-Stained Protein Ladder (-20°C) in the 1st lane.

> Lane 1: Spectra Broad Range Ladder (5 µl)

> Lane 2: Sample 1 (20 µl)

> etc…

1	2	3	4	5	6	7	8	9	10
Ladder		S1	S2	S3	S4	S1	S2	S3	S4

(11) Run **60 V for 5 min**, then run **120 V for 50 min**. Check the gel every 15 min until the sample line reaches the black line.

(12) In mean time get together transfer chamber, blotting pads, filter paper, PVDF membranes, and tubs for soaking everything.

(13) Soak blotting pads and chamber in transfer buffer in tub.

(14) About 10min before gel is done. Rinse PVDF in methanol, rinse, soak in transfer buffer.

(15) When gel is done running. Rinse off gel in DI. Pry open and cut top and bottom off with razor blade. Gels are delicate! Soak filter paper in transfer buffer and then lay on top of gel, being carefully to push out bubbles. Flip and put membrane on top of gel.

(16) Assemble layers with membrane facing toward red (anode, +). Put **ice pad** in chamber. Fill chamber with **1X transfer buffer** (need a little more than 1L) and run at **90V for 90min**.

(17) Block **1hr at RT** with shaking in blocking buffer (LI-COR Odyssey Blocking Buffer (TBS))

(18) Prepare primary antibody solution. Use blocking buffer to make the antibody solution to **1:100**.

> Ex: 1 ml AB = 1 ml Blocking Buffer + 10 µl antibody

(19) (If applied) Prepare control group primary antibody solution. Same as above.

(20) Incubate between Parafilm overnight at **4°C** (save Ab solution and add sodium azide to 0.01%) with primary antibody solution.

(21) Wash in TBST for **5 min, 3X.**

> TBST = TBS + Tween 20 (0.1% v/v Tween 20)

> Ex: 100 ml TBST = 100 μl Tween 20 + 100 ml 1X TBS

(22) (If applied) Incubate between Parafilm overnight at 4°C (save Ab solution and add sodium azide to 0.01%) with control primary antibody solution.

(23) (If applied) Wash in TBST for 5 min, 3X.

(24) Prepare secondary antibody solution. Use blocking buffer to make antibody solution to **1:5000**.

> Ex: 10 ml secondary AB = 10 ml Blocking Buffer + 2 μl secondary antibody

(25) Incubate secondaries for **60 min at RT** with shaking in dark.

(26) Wash in TBST for 5 min, 2X; then wash in TBS for 5 min, 1X.

(27) (If applied) Incubate secondaries for 60 min at RT with shaking in dark.

(28) Wash in TBST for 5 min, 2X; then wash in TBS for 5 min, 1X.

(29) Keep the membrane in TBS.

(30) Image the membrane.

(31) Membrane can be store in 1X TBS in dark at 4°C for up to 3 days.

B.1.7 Sodium phosphate buffer and pH measurement protocol

The sodium phosphate buffer we used in this book is 0.1 M sodium phosphate buffer with pH 7.4. To make a volume of 1 L buffer, add **3.1 g** of **$NaH_2PO_4 \cdot H_2O$** and **10.9 g** of

Na₂HPO₄ (anhydrous) to distilled H_2O. The pH of the final solution will be 7.4. This buffer can be stored for up to 1 month at 4°C.

We used VWR sympHony Electrochemistry B30PCI Benchtop Meters to measure pH of our reagents and buffers. Details can be found at https://us.vwr.com/assetsvc/asset/en_US/id/1094 7627/contents.

Calibrate the pH meter:

(1) Turn on the meter, push 'Calibration', and choose pH probe.

(2) Prepare 3 fresh buffers with pH = 4, 7, and 10 in separate containers.

(3) Rinse the probe with deionized water. Blot dry with a lint-free cloth.

(4) Put the probe in one buffer and push 'Read'. Wait until the display is stable.

(5) When the reading is stable, the meter prompts for the next calibration point. Do steps 3-4 again for additional buffers. The calibration is complete when the last buffer is read.

Measure pH of samples:

(1) Turn on the meter.

(2) Rinse the probe with deionized water. Blot dry with a lint-free cloth.

(3) Put the probe in the sample and push 'Read'. Wait until the display is stable.

(4) When measurements are done, storage the probe in its storage liquid.

* **Note 1:** The calibration and sample measurement conditions such as temperature should be as similar as possible.

* **Note 2:** During measurements, make sure that there are no air bubbles under the probe tip. Gently shake the probe from side to side to remove any air bubbles.

* **Note 3:** Make sure that the reference junction is fully in the solution.

* **Note 4:** Do not put the probe on the bottom or sides of the container since it is very sensitive and fragile.

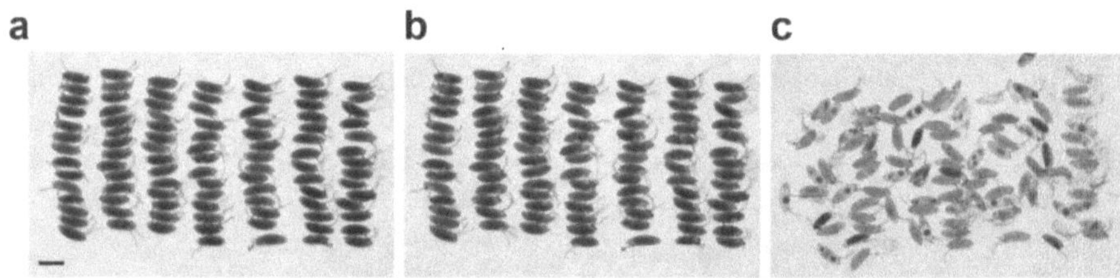

Figure B1. *Drosophila* embryo assay by stereoscope imaging. Bright-field images from time-lapse movies of approximate 100 *Drosophila* embryos mounted in halocarbon oil and lined up on apple juice agar plates (a) before and (b) after body axis elongation, and (c) after hatching. Tissue elongation situation (fully-elongated, partially-elongated, not-elongated, and no celluarization) and fraction of hatched embryos were quantified manually through the movies. E-cadherin is over-expressed in these embryos resulting in a large portion of embryos with defected tissue elongation (b) and defected hatching (c). Error bar, 500 μm.

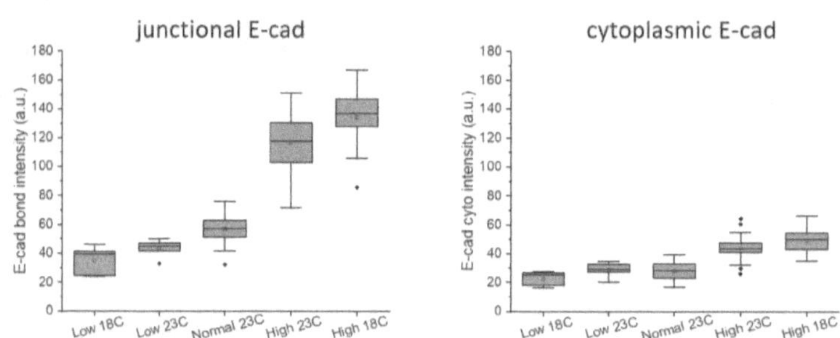

Figure B2. E-cadherin levels are systematically modulated. (a) Junctional and (b) cytoplasmic E-cadherin levels in the germband tissue from embryos with different genotypes are quantified in fixed samples following methods in Section 3.2.6. n = 7-41 fixed embryos. E-cadherin levels in Chapter 3 are differences between junctional and cytoplasmic E-cadherin intensities (Fig. 3.2b).

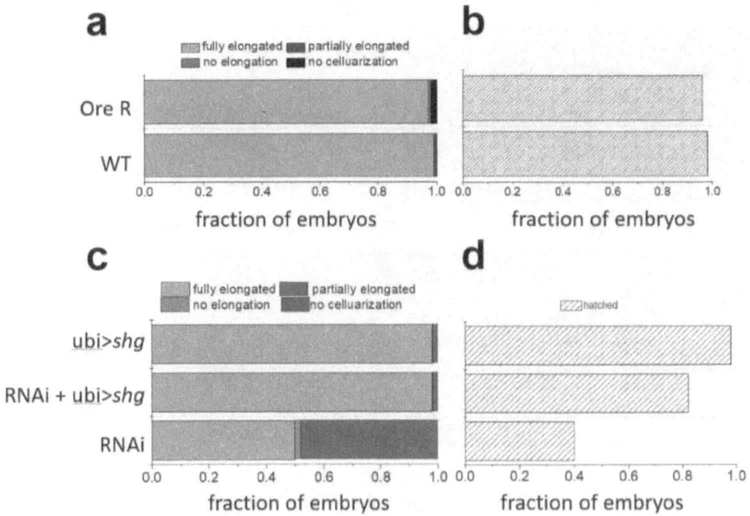

Figure B3. Embryo assay of the control and the rescue groups. (a) and (b) Confirmation of the control group WT by embryo assay. In the control embryos (WT), their endogenous *shg* is replaced by shg:GFP. Their germband tissue elongation (a) and embryo hatching (b) are the same with the wild-type *Oregon R* flies, confirming the control group (WT) set in this study is wild-type. (c) and (d) Defects in embryos with knocked-down E-cadherin by RNAi are rescued by adding in extra E-cadherin. The defected germband tissue elongation (c) and embryo hatching (d) in RNAi embryos are largely eliminated by adding in extra E-cadherin through transgene ubi>shg (RNAi + ubi>shg), confirming that the RANi in the study only knocks down E-cadherin levels.

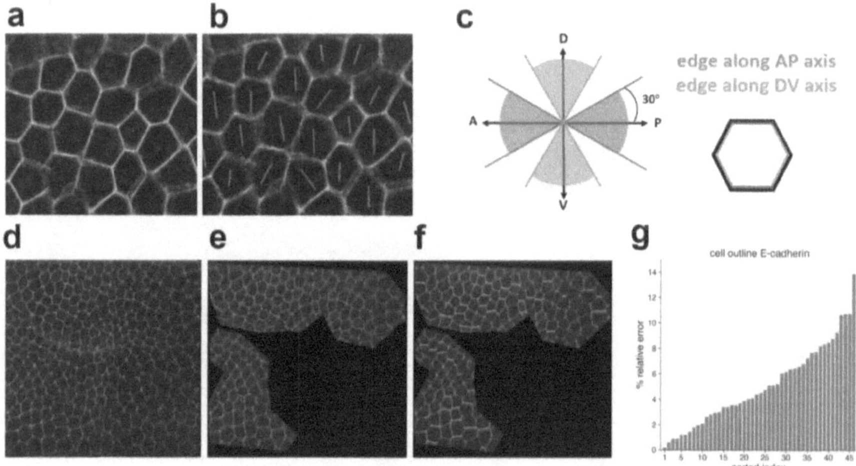

Figure B4. Protein fluorescent intensity measurements. (a) and (b) Manual measurement of protein intensities using ImageJ. 3-pixel-wide straight lines were drawn along cell edges to measure junctional protein intensities (a) and inside of cell to measure cytoplasmic protein intensities (b). (c) Quantification of planar polarity of proteins. Cell edges within 30 degrees away from the anterior-posterior axis are considered as along AP axis (magenta); cell edges within 30 degrees away from the dorsal-ventral axis are considered as along DV axis (cyan). (d)-(g) Automated protein intensity measurement using EPySeg and custom code. (d) Source image recorded in 8-bit grayscale. (e) Segmentation of the cell outline network (green) overlaid on the source image in regions of interest with good projection. (f) Identification of planar polarized cell outlines following standards described in (c). (g) Percent relative errors between manual and automated measurements of mean cell outline E-cadherin intensity.

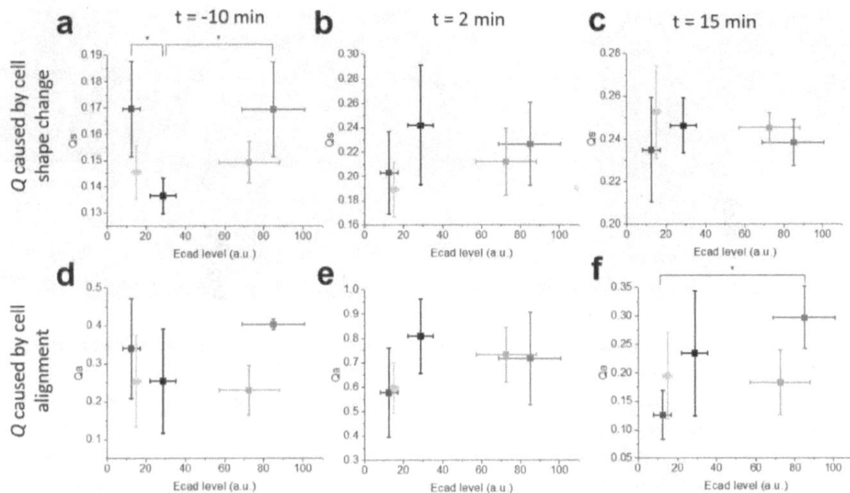

Figure B5. E-cadherin levels modulate cell shape alignment Q in different ways. Cell shape alignment Q describes both cellular shape anisotropy and alignment of cell shapes in the tissue and can be decomposed as $Q = Q_s Q_a$. (cf. Eq. (A8)), where Q_s represents cellular shape anisotropy (a-c) and Q_a represents cell shape alignment (d-f). At 10 minutes before the axis elongation, Q_s shows (a) the same biphasic relationship with E-cadherin levels as p, while (d) Q_a is independent with E-cadherin levels, indicating that at this moment Q is modulated by cell shape change. (b) and (e) At 2 minutes after the onset of axis elongation when Q reaches its peak, Q, Q_s, Q_a all become independent with E-cadherin levels. At 15 minutes after the onset of axis elongation, (c) Similar as p, Q_s shows no relationship with E-cadherin levels, while (f) Q_a increases with higher E-cadherin levels, indicating that at this moment the change of Q is dominated by cell alignment.

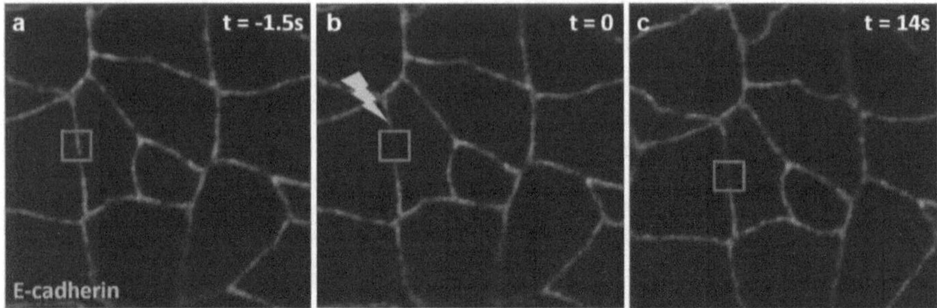

Figure B6. Analysis of E-cadherin dynamics by fluorescence recovery after photobleaching (FRAP). (a)-(c) Live confocal images of E-cadherin from embryos expressing UAS>shg:GFP before (a), during (b), and after (c) photobleaching using 488 nm laser. Red square, ROI covering photobleached and manually tracked area to measure junctional intensity. (a) Before photobleaching, junctional bond intensities in the ROI were measured and averaged to be initial intensity. (b) Photobleaching was applied to the ROI with strong 488 nm laser at $t = 0$. Fluorescence was almost eliminated the ROI. (c) The bleached region of the bond was tracked and the intensity was measured. The intensity recovered over time after photobleaching, but could not recover to the initial level.

Appendix C: *Drosophila* Fly Stocks

Here are the lists of all the fly stocks I generated or requested to work on from 2016 to 2021. Copies of these fly stocks are stored in the Kasza Lab. Numbers starting with '#' (e.g. #32904) are stock numbers in the Bloomington *Drosophila* Stock Center.

C.1 General use

Drosophila balancer:

[1]   Sp/CyO; Dr/TM3,sb

[2]   Dr/TM3,sb (III)

[3]   VgD/CyO (II)

[4]   w, lethal/FM7; +/+; Sb/TM3,ser

Controls:

[5]   Oregon R (wild-type control)

[6]   yw (yellow-body, white-eye)

Drivers:

[7]   mat67; mat15

[8]   w; Sp/CyO; mat15

[9]   mat67; Dr/TM3,sb

[10]   (with slam) mat67/CyO; Dr/TM3,sb

[11]   mat67; +/+

[12]   +/+; mat15

Cell membrane marker:

[13]   +/+; sqh > gap43mCherry ->attP2/TM3,sb

[14]   yw; Sp/CyO; sqh > gap43mCherry ->attP2/TM3,sb (no UAS)

[15]   sqh > gap43mCherry ->attP40/CyO; +/+

[16]   yw; sqh > gap43mCherry ->attP40/CyO; Dr/TM3,sb

C.2 E-cadherin

[17]   ubi > ECad:GFP; Dr/TM3,sb

[18]   #32904 (shg) y sc v; TRiP ->attP2 (III)

[19]   #38207 (shg) y sc v; TRiP ->attP40 (II)

[20]   #58494 w; UAS > shg (II)

[21]   UAS>shg (II); Dr/TM3,sb

[22]   #58366 u; ubi > shg-Venus/CyO

[23]   ubi > shg-Venus/CyO; Dr/TM3,sb

[24]   #58789 yw; shg-mTomato (II)

[25]   #65589 w; UASp > shg ->attP2 (III)

[26]   Sp/CyO; UASp > shg ->attP2 (III)

[27]   #60584 yw; shg-GFP (II)

[28]   shg-GFP/CyO; Dr/TM3,sb

[29]   #59014 w; shg-mCherry (II) (ws)

[30]   #58445 UASp > shg-GFP (III)

[31]   Sp/CyO; UASp > shg-GFP

[32]   (shg) Sp/CyO; TRiP ->attP2 (III)/TM3,sb

[33]   (shg) TRiP/CyO; Dr/TM3,sb

[34]   UAS>shg (II); Dr/TM3,sb

[35]   (shg) sqh > gap43mCherry ->attP40/CyO; TRiP ->attP2 (III)

[36]   (shg) TRiP; sqh > gap43mCherry ->attP2/TM3,sb

[37]   UASp > shg (II)/CyO; sqh > gap43mCherry ->attP2/TM3,sb

[38]   ubi > shg-Venus/CyO; sqh > gap43mCherry ->attP2/TM3,sb

[39]   shg-GFP (II)/CyO; sqh > gap43mCherry ->attP2/TM3,sb

[40]   sqh > gap43mCherry ->attP40/CyO; UASp > shg ->attP2/TM3,sb

[41]   sqh > gap43mCherry ->attP40/CyO; UAS > shg-GFP/TM3,sb

[42]   mat67/CyO; UAS>shgGFP/TM3,sb

[43]   (shg) shgGFP/CyO; TRiP/TM3,sb

[44]   UAS>shg/CyO; mat15/TM3,sb

[45]   Sp/CyO; mat15, UAS>Venus-gap43/TM3,sb

C.3 Myosin II

[46]   sqh > GFP:ZipE470A ->attP2 (III)

[47]   sqh > GFP:ZipE470A ->attP40 (II)

[48]   Sp/CyO; sqh > GFP:ZipE470A ->attP2/TM3,sb

[49]   sqh > GFP:ZipE470A ->attP40/CyO; Dr/TM3,sb

[50]   sqh > gap43mCherry ->attP40/CyO; sqh > GFP:ZipE470A ->attP2

[51]   (with slam) mat67/CyO; sqh > GFP:ZipE470A ->attP2/TM3,sb

[52]   (with slam) mat67/CyO; sqh > GFP:ZipWT ->attP2/TM3,sb

[53]   yw; sqh > gap43mCherry(old) -> attP40; sqh > GFP:ZipWT ->attP2

[54]   w; TRiP (myosin II); sqh > gap43mCherry/TM3,sb

[55]   Sp/CyO; sqh > sqh-mCherry

[56]   mat67/CyO; sqh > sqh-mCherry/TM3,sb

C.4 Other proteins

Actin:

[57]   #35521 Sp/CyO; sqh > moesin-mCherry

[58]   #58988 w; sna/CyO; UASp > F-tractin-tomato/TM2

[59]   #58714 w; Sp/CyO; UASp > lifeAct-RFP

[60]   mat67/CyO; sqh > moesin-mCherry/TM3,sb

[61]   mat67/CyO; UASp > F-tractin-tomato /TM3,sb

[62]   mat67/CyO; UASp > lifeAct-RFP /TM3,sb

[63]   shgGFP/CyO; sqh > moesin-mCherry /TM3,sb

[64]   shgGFP/CyO; UASp > F-tractin-tomato /TM3,sb

[65]   shgGFP/CyO; UASp > lifeAct-RFP /TM3,sb

Patterning genes:

[66]   117 (Resille:GFP); bnt/TM3,sb

[67]   ubi > DECad:GFP; bnt/TM3,sb

[68]   ubi > DECad, sqh > sqh-mCherry/CyO; bnt/TM3,sb

[69]   sqh:GFP; bnt/TM3,sb

[70]   #3299 cn bw sp sna twi /CyO (II)

[71]   #6147 cn twi bw pr/CyO

[72]   #2311 sna cn bw speck/CyO

[73]   #2381 cn twi bw speck/CyO

[74]   #3630 Diap st kni bcd rn p/Tm3,Sb

Rhodamine pathway:

[75]   w; UASp > VenusRoK[K116A] ->attP40/CyO

[76]   UASp > VenusRoK[K116A]/CyO; Dr/TM3,sb

www.ingramcontent.com/pod-product-compliance
Lightning Source LLC
LaVergne TN
LVHW040017200726
843493LV00005B/1298